The Equatorial Rain Forest:
a geological history

To
J.A.F.

The Equatorial Rain Forest:

a geological history

JOHN R. FLENLEY

University of Hull
and
Australian National University, Canberra

BUTTERWORTHS

LONDON — BOSTON

Sydney—Wellington—Durban—Toronto

THE BUTTERWORTH GROUP

ENGLAND

Butterworth & Co (Publishers) Ltd
London: 88 Kingsway, WC2B 6AB

AUSTRALIA

Butterworths Pty Ltd
Sydney: 586 Pacific Highway, Chatswood, NSW 2067
Also at Melbourne, Brisbane, Adelaide and Perth

SOUTH AFRICA

Butterworth & Co (South Africa) (Pty) Ltd
Durban: 152—154 Gale Street

NEW ZEALAND

Butterworths of New Zealand Ltd
Wellington: T & W Young Building,
77—85 Customhouse Quay 1, CPO Box 472

CANADA

Butterworth & Co (Canada) Ltd
Toronto: 2265 Midland Avenue, Scarborough,
Ontario M1P 4S1

USA

Butterworths (Publishers) Inc
Boston: 10 Tower Office Park, Woburn, Mass. 01801

First published 1979

©
Butterworth & Co (Publishers) Ltd
1979

ISBN 0 408 71305 4

British Library Cataloguing in Publication Data

Flenley, John Roger
  The equatorial rain forest.
  1. Forest flora 2. Rain forests 3. Paleobotony
  I. Title
  581.9'09'152   QK938.F6   78-40362

  ISBN 0-408-71305-4

Typeset by Butterworths Litho Preparation Department
Printed in England by Page Bros (Norwich) Ltd.

Preface

Twenty years ago the rain forest was regarded as essentially static, a museum piece, a survival from far into the geological past. The ice ages of temperate regions were thought to have left the tropics substantially unaffected, or to have been reflected there as 'pluvials'. The supposed stability of the rain forest was used to bolster the theory that floristic diversity endowed stability, and the diversity was itself explained in terms of the environmental stability.

Recent research, much of it in the last five years, has shown how wrong were these ideas. The equatorial environment is now believed to have changed markedly in the past, and ice ages in temperate areas were, on the whole, times of aridity in the tropics, not pluvials. The rain forest, and other vegetation of equatorial regions, is now shown to be in a state of considerable instability. The effects of the last major climatic change are still reverberating there, particularly in montane areas.

We have also had to revise our ideas about successional changes in equatorial vegetation. Successions no longer seem to fit into the rigid moulds previously provided for them, and the individualistic behaviour of species is being more widely recognised.

The duration and extent of man's influence on equatorial vegetation has previously been seriously underestimated, but evidence is rapidly accumulating to correct this error.

All these factors encouraged me to write this book. So many decisions are being made at present which will drastically affect the future of equatorial vegetation that I felt it important the facts should be readily available. I hope also that this book will influence a new generation of students, especially in equatorial countries, to study equatorial vegetation as a dynamic entity with a varied and highly significant history.

Despite its title, this is not a work dedicated only to the study of the geological history of the equatorial rain forest. It also covers the other vegetation types of equatorial regions, and it treats vegetational history not in a purely geological manner, but also from a palaeoecological viewpoint. It is in fact an attempt at a vegetational palaeoecology of equatorial regions.

I could not have written this book without the help and forbearance of my wife Jill and my daughters Eleanor, Frances and Yvonne. I thank Professors D. Walker, H. R. Wilkinson and J. A. Patmore for providing facilities for my work. I wish also to thank Donald Walker for critically reading the whole text, and the following for helpful criticism and/or assistance: K. J. Ackermann, M. Arney, D. Aunela, E. S. Barghoorn, R. R. Dean, S. E. Garrett-Jones, J. Golson, M. Gray, J. C. Guppy, D. Guy-Ohlson, T. van der Hammen, A. J. Hanson, G. S. Hope, M. Horgan, C. A. Joyce, A. P. Kershaw, A. Key, J. G. Lindsay, D. Livingstone, R. J. Morley, J. Muller, J. Ogden, C. D. Ollier, J. S. Pethick, J. M. Powell, M. L. Salgado-Labouriau, H. Salzmann, K. Scurr, G. Singh, A. P. Vayda, N. M. Wace, D. Watts and W. Wilkinson. I am also grateful to those many authors who have permitted the use of their illustrations, which are acknowledged beneath each one. I also thank my publishers, Messrs Butterworths, for their friendly patience and for the award of the Butterworth Scientific Fellowship, 1974, which encouraged me to write and made it financially possible.

J. R. F.
Canberra

Contents

1
Present Vegetation and its Biogeographical Problems

1.1 INTRODUCTION

Anyone who has circled the earth in a satellite equipped for time travel need read no further; he will already have seen whole continents migrate and collide, ice sheets expand and contract, sea levels rise and fall, floras evolve and become extinct. The ordinary student, however, must discover the history of vegetation by the more difficult but more satisfying methods employed by Sherlock Holmes, in other words by the collection and assessment of evidence. Nowhere in the world is this study more worthwhile than in the equatorial regions, for it is there that vegetation is at its most complex, previous knowledge at its minimum, and problems about the ecological status and future of vegetation most urgent.

For the purposes of this book, I have considered the regions within about 10° latitude north and south of the equator, which I describe as 'equatorial'. There is no particular theoretical justification for this limitation, nor have I applied it very rigidly; indeed evidence from the whole of the tropics (i.e. up to 23½° latitude north and south) and even outside this limit has been mentioned where it seems relevant.

Owing to the major movement of the continents in geological time, the pieces of land between 10°N and 10°S have not always been the same pieces. Precisely which continental areas were equatorial at any time, however, is still a matter of some discussion. This book therefore makes no attempt to cover the vegetational history of all these areas, but only of those which are *now* equatorial, wherever they may have been in the past.

There is abundant geological evidence that vegetation we currently regard as equatorial formerly occurred in what are now temperate regions. This is not only the result of continental drift, for in certain periods, such as the Miocene, much of the present temperate regions, as well as the present tropical regions, appears to have borne 'equatorial' vegetation. Climatic change must, almost certainly, be involved here. Although I will mention such occurrences, I will make no real attempt to deal with any vegetation outside equatorial regions.

Twenty years ago scientists regarded the rain forest and most other vegetation of equatorial regions as essentially static, a museum piece, a survival from far into the geological past. They thought ice ages of temperate regions had left equatorial regions substantially unaffected, or had been reflected there as 'pluvials'. They used the supposed stability of the rain forest to bolster the theory that floristic diversity endowed stability, and the diversity itself they explained in terms of the environmental stability.

The aim of this book is to show that all these ideas were wrong. The equatorial environment is now believed to have changed markedly in the past, and I shall present evidence that ice ages were times of aridity in the tropics, not pluvials. In fact the old pluvial theory must be completely abandoned. I shall show that equatorial vegetation, even perhaps the lowland rain forest, has changed dramatically in the geologically recent past, and that the effects of the Pleistocene are still reverberating in this vegetation. I shall conclude that equatorial vegetation is essentially dynamic.

1.2 THE PRESENT VEGETATION OF EQUATORIAL REGIONS

History books usually begin, reasonably enough, with history. Why should a book on vegetational history not begin in the same way? The reasons for giving first a brief résumé of present vegetation are threefold. Firstly, the reader may be rather unfamiliar with present-day equatorial vegetation. Secondly, the evidence available to the vegetational historian is much less abundant and clear than that usually used by political, social or economic historians. This means that constant reference to present-day vegetation is necessary for the interpretation of the evidence. Thirdly, a short discussion of present vegetation will permit the recognition of those areas of study where the historical method has the greatest amount to contribute to the interpretation of present vegetation.

This introduction to the vegetation of equatorial regions is necessarily abbreviated and over-generalised, and for more detail the reader is referred to works such as those of Richards (1964), Baur (1964), Walter (1973), Meggers *et al.* (1973), Lind and Morrison (1974), Longman and Jenik (1974) and Whitmore (1975). For simplicity, vegetation is considered under six headings: rain forest, semi-evergreen forest, savannah, swamps, mountain vegetation and secondary vegetation.

1.2.1 RAIN FOREST

This vegetation type, which perhaps covers more area than any other in equatorial regions, is to be identified with the *jungle* of many popular authors, although the usual conception of jungle involves the idea of a tangle of lianes more likely to be found in secondary forest. In fact *jungle* comes from the Sanskrit word *jangala*, meaning desert (Ollier, 1974).

The term rain forest has been established in the literature for so long (at least since Schimper, 1903), that it cannot easily be replaced, although it is in many respects an unfortunate term. It implies a relationship with rainy climate which is, of course, generally correct. This is, however, not always so; in Australia and other areas, strips of 'rain forest' occur in the humid environment alongside rivers in areas where there is a pronounced dry season. The term is also unreasonably broad. For example, there are areas which bear a forest vegetation in cool everwet regions, e.g. the west coast of North America. Some people have felt justified in calling this a rain forest; yet it bears so little relationship to its equatorial counterpart that the name does more to confuse than to enlighten. There is, however, no concise alternative term, so we shall continue to use it.

Rain forest has often been divided into tropical, sub-tropical and temperate rain forest. Again, these terms are unfortunate, because they are environmental terms, rather than properties of the forest itself. This results in anomalies; for instance several authors describe the occurrence of tropical rain forest in sub-tropical regions. This problem need not unduly concern us, however, since all the rain forest within 10° of the equator may be taken to be 'tropical'. To save needless repetition the word 'tropical' in front of 'rain forest' will be omitted.

Rain forest was formerly found in everwet lowland areas almost throughout the equatorial regions. The three principal units there, the American, African and Indo-Malesian* formations, are distinguished more by their floristic differences than by structural divergence. The forest is usually tall (30 m or more) and contains mature trees of many different heights; often there appears to be a fairly distinct 'canopy' with 'emergents' poking through it. Some authors have described other 'strata' below the canopy, but finding these appears to depend largely on subjective choice of site, so that their objective existence is not substantiated. Epiphytes and lianes are often fairly abundant, as are palms and tree ferns. Shrubs and a ground flora of herbs are sometimes present, although the latter may be sparse. Some equatorial herbs are extremely large, e.g. bamboos and many members of the Zingiberaceae. Bryophytes are present, but less abundant than might be expected in the moist environment.

Although many of the trees look alike and have similar entire laurel-like leaves with 'drip-tips', in fact almost every tree found by an observer in the forest turns out to be a different species, for this is the most diverse vegetation type known. Even for large trees alone, the species area curve rises rapidly and fails to flatten off (*Figure 1.1*). Presumably if saplings, lianes, epiphytes, herbs, bryophytes, etc. were included then the curve would rise even more rapidly, although it might eventually flatten more convincingly.

The plant families which occur in rain forest are very numerous but some of the principal ones are: Euphorbiaceae, Leguminosae, Myrtaceae, Burseraceae, Lauraceae, Myristicaceae, Anacardiaceae and Annonaceae. There is considerable regional variation; for instance the Dipterocarpaceae, which are dominant in South-East Asia, are almost absent elsewhere. The above list refers to trees only, but there are also many important herbaceous families such as Orchidaceae.

The taxonomic diversity of the rain forest led to the failure of attempts to classify it by traditional phytosociological methods (e.g. Braun-Blanquet, 1932), and not until the application of statistical methods (Ashton, 1964), and later computer methods (Webb *et al.*, 1967b,c) were satisfactory classifications or ordinations reached. Even now, most of these can only be related to environment in a most generalised way, usually to soil and moisture factors which are not fully independent of each other.

The rain forest has been subjected to shifting cultivation in many areas, and in Africa much forest

*Indo-Malesia: the Indo-Malesian floristic sub-kingdom of Good (1947), i.e. approximately the political states of India, Bangladesh, Burma, Sri Lanka, Thailand, Laos, Cambodia, Vietnam, Taiwan, Malaysia, Indonesia, the Philippines, and Papua New Guinea. The latinised spelling is adopted to avoid confusion with the political state of Malaysia.

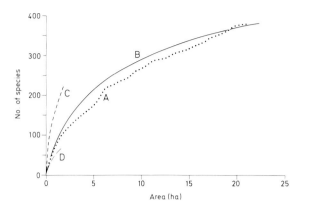

Figure 1.1 Species/area curves from rain forest (lowland Dipterocarp forest) in Malaya.

A. *Additive species/area curve for trees over 28 cm diameter, Jenka Forest Reserve, Malaya.*

B. *Curve calculated from the Index of Diversity of the community in Jenka Forest Reserve.*

C. *Curve for trees over 10 cm diameter, Sungei Menyala, Malaya.*

D. *Curve for trees over 28 cm diameter, Sungei Menyala, Malaya.*

The curves fail to flatten off even at large areas. (After Poore, 1964)

previously thought to be 'primary' or virgin, is now regarded as secondary (Richards, 1955). Complete forest clearance has occurred in so many equatorial regions that it is often impossible to reconstruct the original vegetational boundaries from the surviving relict patches with any degree of certainty. Forest clearance is associated with many problems which will be considered in Chapters 6 and 7.

1.2.2   SEMI-EVERGREEN AND DECIDUOUS FORESTS

These are forests that contain a proportion (sometimes 30—50%; Richards, 1964) of deciduous and semi-deciduous species. They commonly have far fewer total species than the rain forest, and sometimes a single species may dominate. They also tend to lack epiphytes and bryophytes. These forests grow in areas of pronounced wet and dry seasons, and they are therefore sometimes termed monsoon forests, although this is a misleading term since it assumes the correlation with climate which may not always apply. Semi-evergreen forests are most abundant on either side of the equator, rather than at the equator itself. The principal areas in which they are found are Thailand and Burma (especially the famous forests of teak, *Tectona grandis*); East Java (again teak forests); the Lesser Sunda Isles and southern New Guinea; in East Africa and around the margins of the rain forest areas in West Africa; and around the fringes of the Amazon rain forest in Brazil, Guyana, Venezuela, Surinam and the West Indies.

The semi-deciduous or deciduous nature of the vegetation is of considerable interest. It is possible that the deciduous habit arose in response to seasonal drought in the sub-tropics, and later became an adaptation to seasonal cold in temperate regions. Clearly the habit must have arisen many times in different taxa, so there is no reason why the cause of its origin should always have been the same.

Shifting cultivation has also been very widely practised in areas of semi-evergreen and deciduous forest.

1.2.3   SAVANNAH

Broadly speaking, savannah is a grass-dominated vegetation, with or without scattered trees and shrubs. The term savannah has been applied so differently by so many ecologists that it is difficult to define at all. It is more than likely that fire, lit by man, is responsible for the maintenance, and even the formation, of many savannahs. There is good geological evidence, however, that some savannahs have existed in tropical regions since before man is known to have evolved, and are therefore a 'natural' vegetation type.

There are vast areas of savannah in South America, especially the cerrados, south of the Amazon rain forest, and the llanos, to its north. In Africa, savannahs are also widespread, especially to the south of the Sahara and in East Africa. In Indo-Malesia savannah is restricted to smaller areas such as the patanas of Sri Lanka, the cogonales of the Philippines, and the kunai of New Guinea.

1.2.4   MOUNTAIN VEGETATION

The rain forest of the lowlands does not give way sharply to mountain vegetation. Indeed the change is so gradual, and occurs at such a wide range of altitudes, that any boundary must be considered arbitrary. It is true that workers in South America (Cuatrecasas, 1958; van der Hammen, 1974), in Africa (Hedberg, 1951; Troll, 1959), and in South-East Asia (van Steenis, 1972), have considered 1000 m as a suitable boundary, but this can be taken as only the vaguest of guides.

The mountain forests are evergreen but less diverse than the lowland forest; they are also usually less tall but richer in epiphytes and tree ferns. Epiphytic bryophytes may be particularly abundant at higher elevations, clothing every branch and trunk to produce the so-called 'mossy forest'. The boundary between the Lower and Upper montane forests is often fairly clear (Whitmore, 1975).

At elevations above about 3400 m, or frequently somewhat lower, the forest is replaced by non-forest vegetation, often dominated by genera with temperate affinities, the so-called stenotherm or microtherm genera (van Steenis, 1934—36). Frequently there is an intermediate zone of shrubs in which the Ericaceae may be prominent. Isolated small trees may be found at higher altitude, so it is best to distinguish between the forest limit, where closed forest ceases, and the tree line, above which no isolated trees occur.

Above the forest limit the vegetation is extremely variable. On some mountains, for example Mt Kinabalu, Borneo, you emerge from the forest onto almost bare rock. More usually there is a herbaceous vegetation in which grasses dominate to a greater or lesser extent.

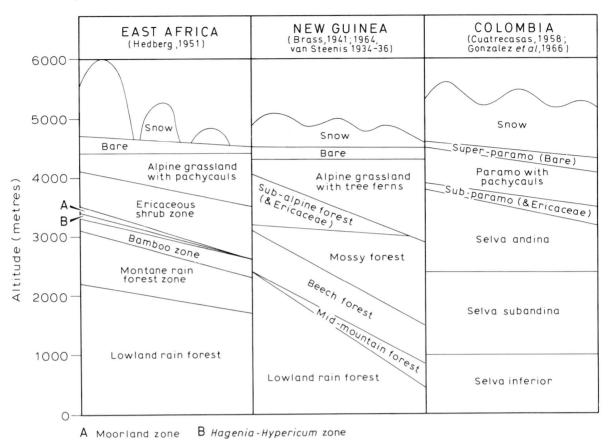

A Moorland zone   B *Hagenia-Hypericum* zone

Figure 1.2 Altitudinal zonations of equatorial vegetation. There is a broad parallelism between zonations in all three regions. (After Flenley, 1967; Troll, 1959 and other authors shown in the diagram)

This has often been called 'alpine' by analogy with the Swiss Alps, but this term suggests a comparison which may not be justified. The term 'tropicalpine' seems less open to objection.

The snow line on equatorial mountains is at about 4500m (Troll, 1959). There is considerable variation in relation to precipitation, and valley glaciers may come well below the regional snowline. Seasonal variation in snow deposition and accumulation is at a minimum, in line with the pronounced lack of seasonality in the climate.

Many attempts have been made to subdivide mountain vegetation on an altitudinal basis, e.g. Cuatrecasas (1958) for South America, Hedberg (1951) for Africa, and van Steenis (1934—36) and Robbins (1958) for South-East Asia. Their results have been summarised in *Figure 1.2*. None of these attempts is entirely satisfactory for a variety of reasons. In the first place, vegetation, if it obeys any laws at all, obeys statistical ones. Any zones, therefore, can only be regarded as areas of high probability for the occurrence of a particular vegetation type (e.g. Walker and Guppy, 1976). This may be illustrated by data showing the number of tree species recorded in plots at various altitudes in New Guinea (*Figure 1.3*). It is clear that at any given altitude this number may take quite a range of values. It is, of course, equally clear that there is an overall trend of decline in diversity with increasing altitude.

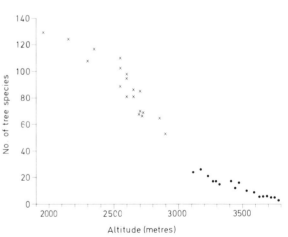

Figure 1.3 The numbers of tree species in plots at various altitudes in New Guinea.
x, Plots of 5000 m² in the Wabag Region (Flenley, 1967; 1969).
•, Plots of 100 m² on Mt Wilhelm (Wade and McVean, 1969).
There is a marked decline in number of species with increasing altitude. The small plots used at high altitude are adequate for the relatively undiverse forest there. (Original)

A second reason for difficulty in zonation is the fact that zone boundaries tend to occur at lower altitudes on small isolated peaks near the sea than on large mountain masses inland. This is the Massenerhebung effect, first noted in the Alps. The effect is particularly marked in South-East Asia (*Figure 1.4*). Among the many explanations advanced for the Massenerhebung are the accumulation of cloud over mountains (Brass, 1941; Grubb and Whitmore, 1966), the effect of wind near the sea (Beard, 1946), the excessive decline in soil fertility with altitude on small isolated peaks (Grubb, 1971), and the differing temperature lapse rate over large mountain masses (Hastenrath, 1968).

It is very desirable that lapse rates should be measured in more places. Few people have the time to take a year's observations and fewer still to take several years', yet variation from year to year may be important. It is to be hoped that the more general availability of automatic apparatus will eventually yield more data. Among the simplest of such devices is the thermal cell developed by Ambrose (1976). Failing all else, those

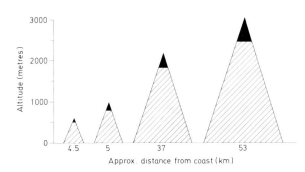

Figure 1.4 The Massenerhebung (mass elevation) effect illustrated by the occurrence of mossy forest on mountains in Indonesia. From left to right: Mt Tinggi (Bawean), Mt Ranai (Natuna Is.), Mt Salak (W. Java) and Mt Pangerango (W. Java). The small mountains near the sea have mossy forest at lower altitudes than the large mountains inland. (After van Steenis, 1972)

on short visits to mountains could do worse than measure the soil temperature at depth. As seasonal variation is slight, and diurnal variation almost eliminated below 30 cm, a probe inserted to 1 m depth gives an approximation to mean annual temperature (Mohr and van Baren, 1960; Schulz, 1960). Preliminary results suggest different lapse rates on different mountains in western Malesia* (Morley, 1976).

Clearly the question 'why do zonations differ on different mountains?' invites the question 'why is vegetation zoned altitudinally at all?'. In answer to the second question there is general agreement, but no proof, that the primary control is temperature. In some cases direct damage by frost affirms the importance of temperature (van Steenis, 1968). It is also true

*Malesia: The Malaysian floristic region of Good (1947), i.e. approximately the political states of Malaysia, Indonesia, the Philippines and Papua New Guinea. The latinised spelling is adopted to avoid confusion with the political state of Malaysia.

that a few transplantation experiments have shown that species rarely flourish outside their ordinary altitudinal range (van Steenis, 1962a), but this could be due to other factors which vary with altitude such as ultraviolet radiation. Phytotron experiments are what is really needed. Even if, as does seem likely, temperature is the causal factor, precisely how it operates is not clear. Brunig (1971) has shown that physiological drought may be likely at high altitudes, due to diurnal variations in leaf temperature, while Grubb (1974) has shown that transpiration may be impaired due to the foggy conditions prevalent on mountains.

1.2.5 SWAMP VEGETATION

Equatorial swamps come in two main kinds: salt and brackish water (mangrove) swamps, and freshwater swamps. There is, of course, no real dividing line between the two, and the vegetation forms a continuum.

Mangrove swamps are vegetated by woody species (mangroves) of several plant genera, for example *Rhizophora*, *Avicennia* and *Bruguiera*, all of which can withstand degrees of inundation by tidal water. The most well-known growth form is that in which abundant aerial roots emerge from the trunk, making an almost impenetrable barrier. Although mangroves may occur on a variety of substrata including coral, they are usually found on silty and clayey deposits. Mangrove propagules are, not surprisingly, adapted to marine dispersal; for example, in *Rhizophora* the seed germinates while still on the tree to produce a vastly enlarged radicle up to 30 cm long, and capable of supporting the seedling in sea-water.

Mangrove swamps are zoned approximately according to the amount of inundation, the salinity of the water, exposure and other factors; the situation is somewhat analogous to the zonation of a temperate salt marsh. A good example is that from the west coast of West Malaysia (Watson 1928).

Freshwater swamps bear vegetation which varies according to altitude, water depth, nutrient status and other factors. Water which is regularly over 2 m deep normally bears either submerged aquatics, or a *schwingmoor* (floating mat), frequently with Cyperaceae as the main structural basis. Shallow water may be dominated by Cyperaceae or Gramineae or by trees and shrubs. Swamp forest may be present in shallow water or where the water table fluctuates. Vast areas in the Amazon Basin, the Irrawaddy floodplain and the lowlands of Borneo, Sumatra, and New Guinea, support swamp forests of various kinds.

A remarkable development found particularly in South-East Asia is the peat-swamp. These gently domed areas of peat may be many kilometres across, and are analogous to the raised bogs of temperate regions (Anderson, 1963, 1964). They are entirely forested, but there is a steady decline in diversity, from over 50 species at the edge, to only 20–30 species in the centre (Anderson and Muller, 1975). Indeed single-species dominance (as by *Shorea albida*) is common.

The successional relationships of these many types of swamp vegetation will receive attention in Chapter 6.

1.2.6 SECONDARY VEGETATION

Most of the vegetation types so far described may be regarded as primary, in the sense that they are believed to be part of the original pattern of vegetation types before man began to affect vegetation. But there is another whole suite of vegetation types which are believed to result directly or indirectly from human activities and are therefore secondary. The term anthropogenic is sometimes applied to such vegetation, but as this literally means 'man-producing' rather than 'man-produced' it is an unfortunate term and best replaced by 'man-made'. The impact of man in equatorial regions as in temperate regions appears to have been principally through felling, burning and agriculture (Thomas *et al.*, 1956). The effect of various minor activities in the forests, such as hunting and the gathering of food and lianes, has remained largely unassessed.

The vegetation of plots abandoned after clearing falls into a sequence of stages which vary enormously depending on the proximity of forest, the duration of clearance, the type of cropping and other factors. The act of clearance exposes the soil to the sun, resulting in direct oxidation of organic matter. If fire is used in the clearance further loss of organic matter results. Most of the nutrients in a tropical land ecosystem are stored not, as in temperate systems, in the soil, but in the biomass, principally the vegetation. Burning results in the gaseous loss of some elements such as nitrogen and the conversion of others to mineral ash which is rapidly leached by the tropical rainfall. Clearance therefore tends to lead to soil degradation and infertility. The result is that primary forest trees are, in many cases, unable to invade cleared land immediately it is abandoned, for their seedlings seem to require higher nutrient status and, in many species, shade. Early dominants of secondary vegetation are therefore usually grasses, tree ferns and small trees. The trees concerned are adapted to the secondary habitat by rapid growth, good seed dispersal, ability to germinate in the open and ability to flourish on poor soils (sometimes the result of nitrogen-fixing root nodules as in *Casuarina* spp. and *Trema* spp.). The secondary trees are considered to have been, before man's activities, 'nomads' in the forest, dependent on rare natural events for their survival (van Steenis, 1958a). Successional aspects of this secondary vegetation will be considered in Chapter 6.

1.2.7 THE GEOGRAPHICAL DISTRIBUTION OF PRESENT VEGETATION

Most vegetation types occur where ecological considerations would lead you to expect them (*Figure 1.5*). Secondary vegetation and agricultural land alone do not follow these relatively simple trends. In general,

man has used the land close to the sea, rivers and other access routes. The latest example is the clearance of land on both sides of the new Amazonian roads. Sometimes geological considerations have been important to man, for instance much of the east coastal plain of W. Malaysia has been cleared during tin-mining operations. The greater clearance of land in lowland Java than in lowland Sumatra is related closely to the much greater population density on Java. Whether land remains in cultivation or reverts to secondary vegetation again depends on complex factors. Soil fertility is important, but economic or other factors may also be very significant. In the New Guinea Highlands, valley bottoms were deserted possibly because malaria was more prevalent there; alternatively they may have been relinquished because they were difficult to defend when population pressure led to increased inter-tribal warfare (Brookfield, 1964).

1.3 THE VALUE OF VEGETATIONAL HISTORY

The study of present vegetation raises a number of problems. Some may be wholly or partially solved by vegetational history. Past vegetation changes may also be evidence of great importance for other sciences such as climatology, geomorphology, stratigraphy, geophysics, archaeology and human geography. Topics to which vegetational history in equatorial regions may have particular relevance are briefly mentioned below.

1.3.1 ANOMALOUS RANGES

The theoretical distribution range for a taxon (a unit of classification such as a species, genus or family) is a circular one resulting from a spread outwards from its centre of origin; the size of the circle would relate to the age of the taxon (Willis, 1922). Distributions fail to reach this theoretical ideal to various extents, and when they depart markedly from it they may be called anomalous. The modes of departure are various, and some are susceptible to historical investigation; a few examples are given below:

Relict ranges
If there are grounds for thinking a taxon to be ancient, for instance if it is taxonomically isolated or morphologically primitive, then it should have a wide range. If it has a restricted distribution, there is a strong possibility that it is a relict and was formerly more widespread. Fossil evidence of this may be readily available if it is a morphologically distinct taxon. Thus the conifer genus *Metasequoia* is restricted now to a single area of China, but occurred also widely in North America and Europe in the Mesozoic and Tertiary Eras.

Disjunct ranges
The range of a taxon is disjunct when the taxon occupies two or more well separated geographical areas. Between these areas barriers of different kinds

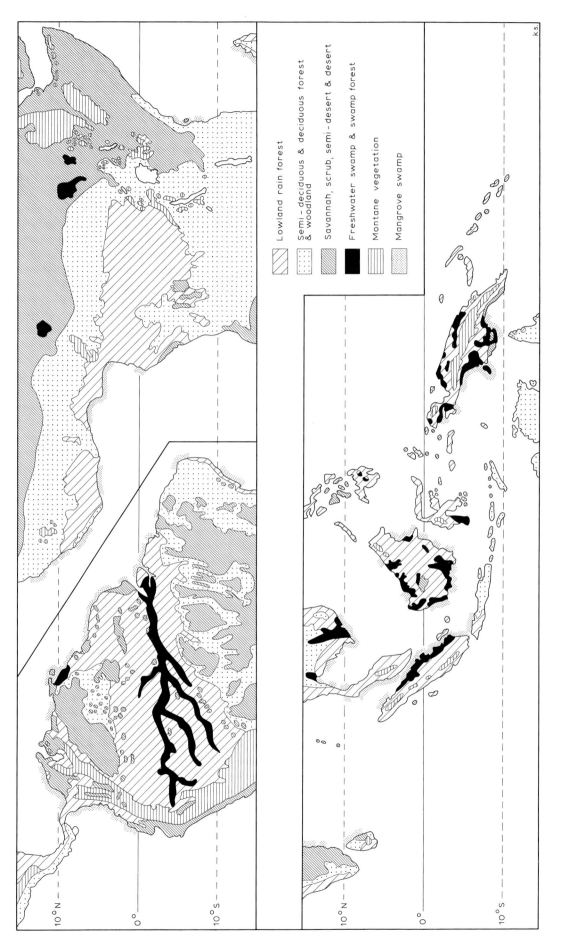

Figure 1.5 The present vegetation of equatorial regions (ignoring recent forest clearance). Miller cylindrical projection. Compiled from various sources including van Steenis (1958b), Clark (1967), Livingstone (1975), van der Hammen (1974), Eden (1974) and Hueck and Seibert (1972)

Lowland rain forest

Semi-deciduous & deciduous forest & woodland

Savannah, scrub, semi-desert & desert

Freshwater swamp & swamp forest

Montane vegetation

Mangrove swamp

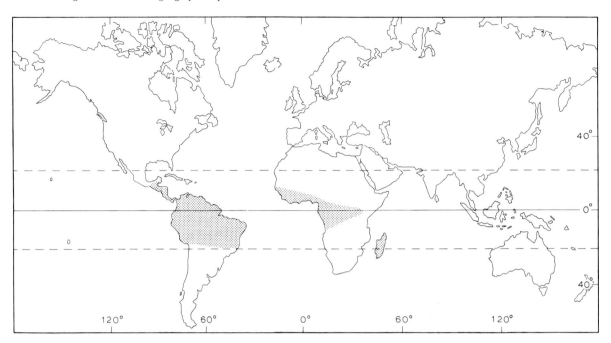

Figure 1.6 Distribution of the genus Symphonia. *A distribution in accordance with plate tectonic theory. (After Good, 1947)*

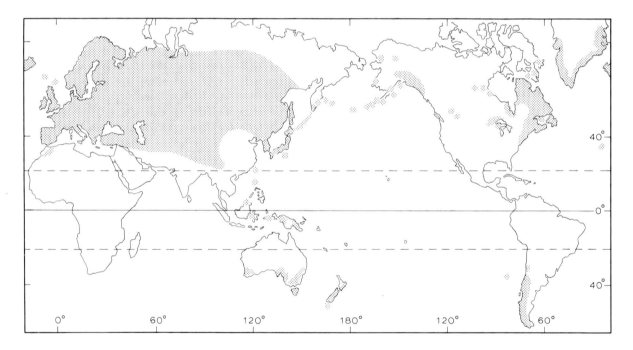

Figure 1.7 Distribution of the genus Euphrasia *showing equatorial occurrences (on mountains) of a mainly temperate genus. (After van Steenis, 1971)*

are normally present. The most common barrier is the ocean, as with the genus *Symphonia* (*Figure 1.6*), that is present in the African and South American tropics, separated by the Atlantic. Another common barrier is a mountain range: for example, differences in plant communities on either side of the Rocky Mountains of North America cannot all be explained by climatic differences. The reverse is the situation for mountain plants, which can be regarded as living on islands in a sea of lowland vegetation. The genus *Euphrasia* is present on numerous high mountains in South-East

Asia (*Figure 1.7*) but totally absent in the intervening lowlands. In the last two cases the barrier is presumably a climatic one: lowland species cannot grow in the mountains and *vice versa*. Another example of climatic disjunction is that exhibited by many taxa in the semi-evergreen forests of South-East Asia (Ashton, 1972). These forests occur in the monsoon climate, with a pronounced dry season, in Thailand and East Java, but are absent from the everwet areas in between (*Figure 1.5*). In some cases land itself is a barrier; the mangrove swamps on both sides of the Isthmus of

Panama show similarities and differences which are probably closely related to their history.

The contribution of historical data in these cases could well be crucial. In the case of major disjunctions between continents, a demonstration that the taxon had a long history on both sides of the barrier might well, in conjunction with geological evidence for continental drift, solve the problem completely. In the case of mountain barriers it would again be relevant to know the history of the disjunct taxon in relation to the age of the mountain range. Did the plant, for instance, exist on both sides before the orogeny, or did it succeed in crossing the barrier in some way?

The problem of mountain floras has long been of interest to biogeographers. In temperate regions it has been shown that historical evidence is decisive; for example in Britain the Late-glacial lowland flora included many Arctic-alpine species which became restricted to mountains during the Post-glacial forest period (Godwin, 1956). In the tropics close relationships between the floras of separate peaks have been established in Africa (Hedberg, 1964) and in South-East Asia distinct 'migration tracks' have been recognised (van Steenis, 1934–36). In South America the various cordilleras of the Andes form the clear migration routes. All tropical montane floras have distinct relationships with temperate floras. Darwin (1859) clearly enunciated the possibility that a former cooling of climate had been very important in aiding the migration of species to tropical peaks from temperate areas, and the idea was thoroughly espoused by Wallace (1869). Not all later workers have agreed with this. Van Steenis (1934–36) for example, preferred the possibility that former continuous mountain chains had existed, although there is little geological evidence for this. The body of data relevant to the tropical mountain floras is now considerable (*see* Chapters 3–5).

In the case of disjunction in the lowlands between two blocks of rain forest, historical evidence could possibly show the former similarity of the now separated blocks.

Vicarious ranges
When two closely related taxa occupy the same ecological niche in different geographical regions, they may be termed vicarious (Good, 1947). There is always a strong supposition that such taxa have a common evolutionary origin and have diverged following isolation. It would be desirable to check this hypothesis against the fossil record, although this does not yet appear to have been done successfully.

1.3.2 SUCCESSIONAL CHANGES

The classical accounts of successions (Clements, 1916; Tansley, 1920) are all related to temperate examples. Tropical successions remain little studied, although they are clearly very different from temperate ones. Even in temperate areas it is impossible to obtain certain knowledge of the precise sequence of events without historical information. It is very tempting to examine, say, a series of concentric zones in the hydroseral vegetation around a lake, and to conclude that the zones also represent successive stages in a temporal sequence. Unfortunately this may be an incorrect conclusion, and it is necessary to examine fossils in the lake deposit before the actual succession is revealed. This technique has led to a critical re-evaluation of many earlier ideas about the temperate hydrosere (Walker, 1970b). In the case of a xerosere, the pages of history are not so well preserved; it is necessary to choose a site of sediment accumulation lying within the area of land under study. One particular type of xerosere, the vulcanosere, is quite abundant in some equatorial regions and may be studied by analysis of deposits from crater lakes.

When the direction in which a succession is moving is known it is still very difficult to know the rate of change. This may have important applications, for example the time needed for secondary forest or logged forest to revert to a forest similar to the primary forest is essential knowledge in planning any sensible use of forest resources. Vegetational history may be able to provide this information.

1.3.3 DIVERSITY

Diversity may be defined in several ways, but is usually some measure of the richness in species of an area, a community or a flora. Whatever criteria are used, tropical lowland forests emerge as very diverse compared with those from semi-arid, temperate or montane habitats. Indeed there is, with many exceptions, a general decline in diversity from equator to pole and from sea-level to the snow line. The more one thinks about this the less obvious appear the reasons for it. The number of plant species in any defined area is the resultant of two factors: the rate of acquisition of species by immigration into the area or by speciation, and the rate of extinction or emigration of species once there. A balanced, but dynamic, equilibrium is achieved when these two rates are equal. This was understood by Darwin (1859), and has been put in more mathematical terms by MacArthur and Wilson (1967). In the case of small islands near a continent, immigration may be the main means of acquiring new species, but in the case of large land areas speciation is usually more important than immigration.

The theories which try to explain the incredible variations in diversity throughout the globe may be grouped as shown in *Table 1.1*. It must be emphasised that the theories are not mutually exclusive. It is clear from *Table 1.1* that several of the theories could be tested if sufficient historical data were available. The non-equilibrium hypothesis, for example, requires that all but the most diverse communities should have a still increasing diversity. Similarly the theory that tropical populations are more sedentary should be susceptible of historical investigation, as perhaps should the idea that speciation rates are high in the tropics. The whole concept of the stability of tropical

Table 1.1 AN OUTLINE OF THE BASIC HYPOTHESES CONCERNING SPECIES DIVERSITY, PARTICULARLY THE INCREASED SPECIES DIVERSITY IN THE TROPICS COMPARED TO TEMPERATE AND ARCTIC REGIONS. (AFTER RICKLEFS, 1973)

NONEQUILIBRIUM HYPOTHESIS
    Time — the tropics are older and more stable, hence tropical communities have had more time to develop.
EQUILIBRIUM HYPOTHESES
  I. Speciation rates are higher in the tropics.
     A. Tropical populations are more sedentary, facilitating geographical isolation.
     B. Evolution proceeds faster due to
        1. a larger number of generations per year.
        2. greater productivity, leading to greater turnover of populations, hence increased selection.
        3. greater importance of biological factors in the tropics, thereby enhancing selection.
  II. Extinction rates are lower in the tropics.
     A. Competition is less stringent in the tropics due to
        1. presence of more resources.
        2. increased spatial heterogeneity.
        3. increased control over competing populations exercised by predators.
     B. The tropics provide more stable environments, allowing smaller populations to persist, because
        1. the physical environment is more constant.
        2. biological communities are more completely integrated, thereby enhancing the stability of the ecosystem.

environments (non-equilibrium hypothesis and equilibrium hypothesis IIB in *Table 1.1*) is also capable of investigation by vegetational history.

1.3.4   STABILITY OF COMMUNITIES

A popular recent hypothesis has been that the more diverse a community is, the more stable it will be (MacArthur, 1955). There are some good reasons for believing this: for instance, the populations of a predator and its prey in a simple community often fluctuate wildly, but in a more complex one the predator can use various prey and fluctuations are thus damped down. The hypothesis has often been bolstered, however, by quoting the rain forest as an example of great stability through geological time. If it could be shown that tropical forests, despite their diversity, have changed markedly in distribution in the Quaternary, and are therefore not particularly stable, this would argue against the hypothesis.

1.3.5   THE EFFECT OF MAN ON VEGETATION

Man is an animal and like all animals he affects his environment to some extent, but the effects of man are astounding. Even primitive man at an early level of evolution may have had sufficient knowledge of tools and fire to affect forest cover, and the activities of prehistoric modern man in this direction are well known. Mesolithic cultures have left tree trunks with the marks of the axe which felled them and forest clearance of this age can be detected now in British pollen diagrams. The activities of neolithic cultures are even clearer; forest destruction becomes obvious and even the pollen grains and fruits of the cultivated plants have been recovered from organic deposits and archaeological sites. Agriculture is also indicated by the presence of weeds, recorded as both pollen and seeds. Indeed, it is even possible to conclude, from the assemblages of weeds present, whether a particular agricultural phase was primarily pastoral or arable.

The success of this historical approach in elucidating the origins of farming and of secondary vegetation in temperate regions encourages one to expect much of it in the tropics. This is particularly so in view of the significance of tropical regions for the origin of cultivated plants. In his classic work on this subject, Vavilov (1951) distinguished eleven centres of origin for cultivated plants. Five of these are wholly or partially in the equatorial regions. There is a good chance, therefore, that the vegetational history of equatorial regions holds important evidence regarding the origins of agriculture.

Several equatorial vegetation types, especially savannahs and grasslands, are of possible secondary origin, or at least maintained by man, and there is every possibility that the factual data provided by the historical approach can also throw light on this field of study, which until now has been dominated by speculation.

The effects of man on vegetation will be reviewed in Chapter 7.

1.3.6   CLIMATIC CHANGES

It is sometimes said that if one examines a set of vegetational changes — as recorded, for example, in a pollen diagram — and removes the effects of natural successions, soil maturation and human and animal activity, then what is left must be due to climatic change. This statement needs several caveats, but it adequately expresses an important viewpoint, that is, the reluctance of reputable biogeographers to attribute vegetational changes too readily to climatic change. There is, of course, unimpeachable evidence for climatic change in the Quaternary, especially from temperate regions. From the equatorial regions the evidence is much less strong. I do not intend to review other than vegetational evidence, but it is perhaps desirable to mention it very briefly:

Glaciation
Most tropical mountains over about 3800 m high bear

evidence of Pleistocene glaciation. Deglaciation has in some cases been dated to between 8000 and 15 000 B.P. (Livingstone, 1962; Gonzalez *et al.*, 1966; Hope and Peterson, 1975; Flenley and Morley, 1978).

Lake levels

Several lakes which are centres of internal drainage show evidence of former higher levels, e.g. the surface of Lake Nakuru in Kenya was formerly 180 m higher than at present (Washbourn, 1967).

$^{16}O/^{18}O$ ratios

The measurements made by Emiliani (1955) claimed to show that the tropical Atlantic surface water was 6° C cooler during the late Pleistocene. Unfortunately it now seems likely that these ratios indicate not so much the temperatures as the amount of water incorporated in world ice sheets at the time, which can only be an indirect indication of tropical temperatures (Shackleton, 1967).

Soils and weathering

In some areas the soil is one which is known to form only under a different climatic régime from that obtaining in the area at present, e.g. some lateritic soils of the Lesser Antilles, India, Senegal and Queensland (Mohr and van Baren, 1960).

Geomorphology

Some physical features of the landscape can only form under certain climates, for example, dunes usually require desert conditions. In some cases present climate does not conform with landforms at all; in Amazonia, Tricart (1974) describes dune systems seen, on satellite photographs, in savannah areas where there is no possibility of dune formation today.

Alluviation has been taken to imply certain climatic conditions, but it is difficult to interpret this evidence since high values for suspended solid matter may be characteristic of rivers in both high and low rainfall areas (Douglas, 1967).

Animal distributions

Many animals show distributions only explicable in terms of climatic change, possibly combined with other geological changes. For instance, in the opinion of Simpson (1971), the bird fauna of the Amazon can only be explained if the Amazon rain forest was divided into several separate blocks during the Pleistocene.

Theoretical considerations

Modern theories of climatic change are legion (Flint, 1971). Some of these, such as Milankovitch's theory, would require little or no climatic change in the tropics. Others, such as the theory of variation in solar output, imply considerable world-wide changes, although it is likely that the high albedo of the glaciated temperate regions would have led to exaggerated change there compared with the tropics. In arid regions there is a possibility that the so-called 'pluvials', if they occurred at all, occurred at quite different times depending on the latitude, for if a general cooling led to decreased atmospheric circulation, all climatic belts might move towards the equator. Thus pluvials near the equator would tend to coincide with interglacials, while those further away would be more prone to occur along with glacials. *Figure 1.8* shows several suggested models for climatic change near the equator.

When climate changes it is likely to do so in an extremely complex manner. A lowering of solar output, for example, might result in a lessening of

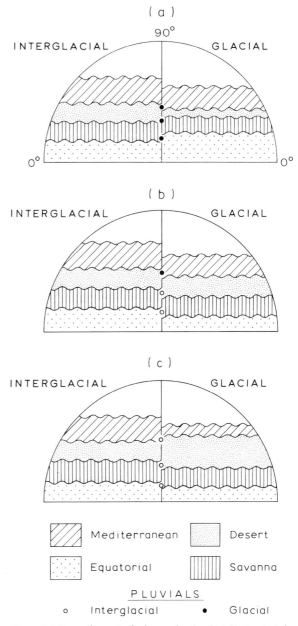

Figure 1.8 Some theoretical schemes for the glacial/interglacial cycle in low and middle latitudes.

(a) The old 'pluvial' theory. All pluvials are synchronous with glacials.

(b) All climatic belts move towards the equator in glacial times. Some pluvials are synchronous with glacials, others with interglacials.

(c) Deserts expand during glacial times. All pluvials are synchronous with interglacials.

Several other such schemes have been proposed. Present evidence favours (c). (Original)

evaporation from the oceans and thus a reduction in precipitation. It could also cause a drop in evaporation from a lake and a rise in lake level, while a reduction in evaporation from land might make the climate moister. At the same time a eustatic fall in sea-level might result in exposure of new land surface and emergence of barriers to ocean currents, both of which could affect climate. The consequent pattern of climatic change would hardly be likely to be simple.

The contribution of vegetational history to the revelation of climatic change has already been immense in temperate regions, and a promising start has now been made near the equator (Chapters 3–5). It is necessary, however, to bear constantly in mind that our interpretations of former vegetation in terms of climate depend absolutely on our understanding of the present relationship between vegetation and climate. Near the equator this understanding is unfortunately all too uncertain.

1.3.7 STRATIGRAPHY

Although the conclusions of vegetational history are not necessarily of direct use in stratigraphy, the primary data — records of assemblages of fossil plants — are of increasing significance. Most parts of the geological column are divided into zones defined according to assemblages of animal macrofossils. Increasingly, however, microfossils are being used: among other advantages is the fact that they may be abundant in small core samples. Pollen and spores are important in this connection for many geological periods, and assume an essential role in the Quaternary. Quaternary stratigraphy cannot rely on the evolution/ extinction cycle which is adequate for other eras, since the time scale is too short. Instead, zonation is based largely on climatic changes (except when absolute dating is available and chronozones are introduced). Since pollen and spores are among the best evidence of climatic change available, their significance will be apparent.

1.4 METHODS OF STUDYING VEGETATIONAL HISTORY

If vegetational history is to be factual, it must be based on fossil evidence. It is true that a study of present floras and vegetation forms a splendid starting point for formulation of historical hypotheses, but these remain untested unless fossil evidence is brought to bear.

The word fossil meant originally anything 'dug up' (Geikie, 1923). It is not necessarily something formerly living for one may speak nowadays of a fossil dune or a fossil lightning strike as well as a fossil bone or leaf. Fossils of living organisms normally consist of those parts resistant to decay, either preserved as they are in life or replaced subsequently by other chemicals. Replacement is by no means universal — even the chemical kerogen, which forms the oldest known

fossils from the Precambrian, is not very different from modern sporopollenin (Brooks and Shaw, 1968). There is therefore no need for the term 'sub-fossil' which sometimes appears in the literature. Indeed as Geikie (1923) says: "The word fossil is sometimes wrongly used as synonymous with 'petrified', and we accordingly find the intolerable barbarism of 'sub-fossil'."

The plant fossils which are the concern of this book may be considered under two headings: macrofossils (leaves, wood, stems, roots, flowers, fruits and seeds) and microfossils (pollen and spores).

1.4.1 MACROFOSSILS

The vast number of macrofossil determinations in the literature is unfortunately no indication of their reliability. In the past it was not uncommon to determine fragments of single leaves or fruits to the species level, and even to erect new species for such inadequate specimens. Often there was not even a microscopic examination of the cuticular structure. The situation in equatorial regions, with enormous floras and many similar leaf and fruit types, was particularly bad. With the exception of fossil conifers, which have been reviewed by Florin (1963) most of these fossils require critical re-examination.

There is one possible approach with fossil leaf floras which may be of value to the vegetational historian. It has been shown for modern vegetation that statistical distributions of leaf characters are reasonably constant for a given major vegetation type, characteristic of a given climatic zone, but differ for other types and zones. The characters frequently used are leaf size-class, serrated/entire margin, and presence/ absence of drip-tips. The proportion of entire-margined leaves is believed to be the best indicator of the mean annual temperature (Wolfe, 1971), as shown in *Figure 1.9*. This idea can be used to suggest whether a fossil leaf flora has affinities with equatorial or sub-tropical modern vegetation for example, without the need to rely on suspect taxonomic determinations.

As with all fossils, problems of selective dispersal, deposition, incorporation, preservation and collection, must be constantly borne in mind when interpreting fossil assemblages.

1.4.2 MICROFOSSILS

The science of palynology (Greek *palynein* – dust), the study of pollen, spores and other microfossils, has taken large strides in recent years. Pollen grains and spores display great diversity of appearance, so that they can frequently be referred, after microscopic examination, to a family and genus, or even to a species. They are produced in large numbers by the parent plants, and often spread widely in air and water. The outer coat, the sporoderm, preserves well in the reducing conditions present in most waterlogged sedi-

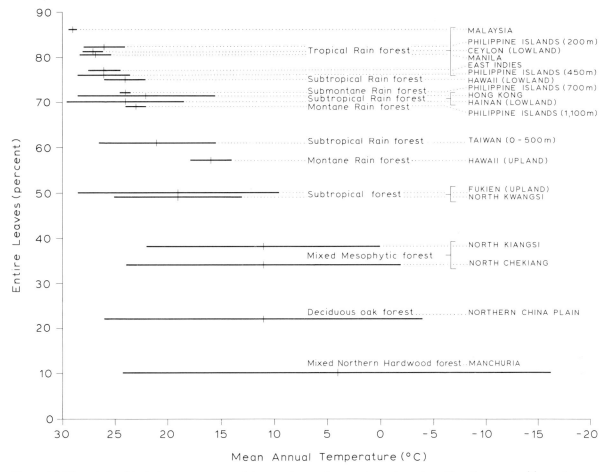

Figure 1.9 The relationship between mean annual temperature and percentage of species with entire-margined leaves in some modern floras. The horizontal lines give an indication of mean annual temperature range, but are plotted equally either side of the mean and therefore do not necessarily indicate mean maxima and minima. High percentage of entire-margined leaves in a fossil flora can be taken to indicate a rain forest flora. (Data from Wolfe, 1971)

ments, and it is the characters of the outer coat which are used in determination. The microfossils can be extracted by a variety of physical and chemical methods, and the relative proportions of each type in a sequence of samples can be shown in pollen and spore diagrams. Sometimes it is desirable to show results as grains per gram of sediment or even, when it is possible to determine sedimentation rate, as grains per cm^2 per annum. Details of all these techniques are given in standard textbooks on palynology, such as those by Faegri and Iversen (1975) or Tschudy and Scott (1969). The steady refinement of its methods has made palynology one of the most important sources of information for the vegetational historian.

All techniques have their disadvantages, however, and palynology is no exception. Some of the problems associated with all fossils are exaggerated in the case of palynology. Widespread dispersal, for instance, although an advantage in leading to preservation of a record of the land flora in sediments, can give misleading results. Thus pollen of *Nothofagus* is recorded on Tristan da Cunha by Hafsten (1960) although the nearest trees of this genus are 4000 km away, in South America. Differential preservation is probably responsible for the fact that some families, for instance

Juncaceae, are scarcely found in the pollen record. Secondary deposition is always likely with such small objects; for example re-deposited Carboniferous spores are frequent in British Late-glacial sediments. Differential production of pollen grains by different species makes the quantitative reconstruction of past vegetation from pollen diagrams a tricky procedure. Despite these and other problems, palynological results display a strong internal consistency which gives some confidence to the interpreter.

Pollen in Quaternary sediments can usually be referred to extant taxa without much difficulty, but in earlier eras this is less easy. Tertiary grains are sometimes placed in general with the ending -*ites* (as *Alnites*) if they closely resemble an extant genus. Cretaceous and Tertiary pollen is also often placed in a form genus such as *Triporites* or *Tricolpopollenites* — names which speak for themselves.

When pollen analysis is used as an indicator of geological history, it can do no better than adhere to geological principles. One of the greatest of these is Lyell's (1850) doctrine of uniformity — 'The present is the key to the past'. In palynology this means that the pollen rain from existing plant communities should be used to interpret fossil pollen spectra in vegetational terms. The principle of uniformity must, however, be applied with care. In the first place it is not sufficient merely to show that a modern pollen spectrum resembles a fossil one, and thence to argue that both were produced by the same type of vegetation. It is also necessary, in theory at least, to show that no other vegetation type produces a similar pollen spectrum. In practice this is impossible, but at least similar vegetation types should be eliminated. Secondly, the fossil spectrum may have come from a vegetation type no longer extant, either because species extinction has occurred or because species have become associated into new vegetation types. In spite of these difficulties, modern pollen rain samples (from moss polsters, water tanks, sticky slides, air samplers, rain gauges, pollen traps or other devices) remain one of the most logical methods for interpreting fossil results.

Palynology in tropical regions had long been advocated, for example by von Post (1946), who recognised several areas as particularly worthy of study. A cautious note was sounded by Faegri (1966) who feared that in the tropical lowlands limited pollen production and dispersal combined with poor preservation and the fantastic diversity of pollen types might make palynology impracticable. These fears have so far proved groundless (Flenley, 1973), although there is still little work completed on the lowlands, unless the pre-Quaternary studies on offshore sediments be included. On mountains in the tropical zone a considerable start on Quaternary palynology has been made, as will appear in Chapters 3–5.

2
The Prelude to the Quaternary

2.1 INTRODUCTION

Since almost all equatorial regions are dominated by angiosperms today, the vegetational history of these regions can conveniently be taken to have begun when angiosperms first appear in substantial numbers in the fossil record. This happened during the Cretaceous. These early flowering plants replaced a vegetation dominated by conifers, cycads and ferns (Axelrod, 1963). The conifers already exhibited a marked separation into northern and southern genera as do most modern conifers.

Florin (1963) explained this isolation of northern and southern floras by means of a barrier to distribution, the Tethys Sea. This was visualised as an epicontinental sea following a line from the Mediterranean through the Caucasus, Himalayas and South-East Asian mountains, with a corresponding feature separating North and South America. Thus he dispensed with the need for continental drift as an explanation. Now, however, from the discoveries of plate tectonics, we know that the Tethys was no mere epicontinental sea, but, at least in some periods, a major ocean, separating the southern 'Gondwanaland' continents from the northern ones of 'Laurasia' (Adams and Ager, 1967); this explains present conifer distributions even more satisfactorily, for a principal component analysis of conifer distributional data produces a result in which the continents take up pre-drift positions on an imaginary globe (*Figure 2.1*) (Sneath, 1967). The floras of these times already showed latitudinal zonations which broadly parallel those of present floras (Barnard, 1973).

2.2 THE CRETACEOUS PERIOD
(136—65 M years ago, Harland *et al.*, 1964)

The southern continents had already begun their long drift northward in the early Mesozoic; this was to lead eventually to the elimination of the Tethys. The possible positions of the equatorial pieces of continental crust in the Middle Cretaceous, about 100 ± 10 M years B.P. are shown in *Figure 2.2*. This will doubtless be revised as more information becomes available from palaeomagnetism and sea-floor dating. One possible minor alternative which has already been suggested (Burton, 1970) is that the Malay peninsula might have been joined to the east side of India in the Mesozoic, although later evidence suggests it was part of Laurasia (McElhinny *et al.*, 1974). If the map is accepted as it stands, the southern continents were still remarkably close together at this time, with the South Atlantic not yet in existence. South-East Asia — or those small parts of it that then existed — appears to have been part of Laurasia. The Tethys was still wide open except between western Europe and Africa where land connections may have existed. It was on this configuration of continents, or something rather like it, that angiosperms first successfully competed with gymnosperms.

The origin of angiosperms is a problem which has exercised many minds over a long period (Hughes, 1976). Several questions may be asked concerning this origin: when, where and why did it happen, what were the precursors of angiosperms, and what were the first angiosperms like? Wolfe *et al.* (1975) conclude that 'there is no unequivocal evidence that indicates a

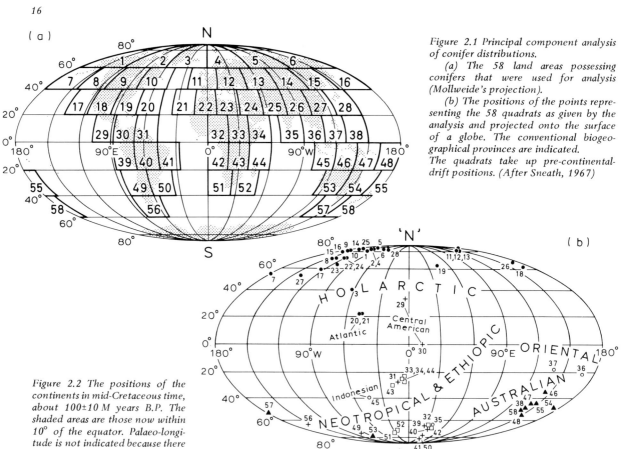

(a)

Figure 2.1 Principal component analysis of conifer distributions.

(a) The 58 land areas possessing conifers that were used for analysis (Mollweide's projection).

(b) The positions of the points representing the 58 quadrats as given by the analysis and projected onto the surface of a globe. The conventional biogeographical provinces are indicated.
The quadrats take up pre-continental-drift positions. (After Sneath, 1967)

(b)

• Holarctic   ○ Oriental   + Ethiopic   □ Neotropical   ▲ Australian

Figure 2.2 The positions of the continents in mid-Cretaceous time, about 100±10 M years B.P. The shaded areas are those now within 10° of the equator. Palaeo-longitude is not indicated because there is no way of measuring it. Mercator projection. (After Smith et al., 1973)

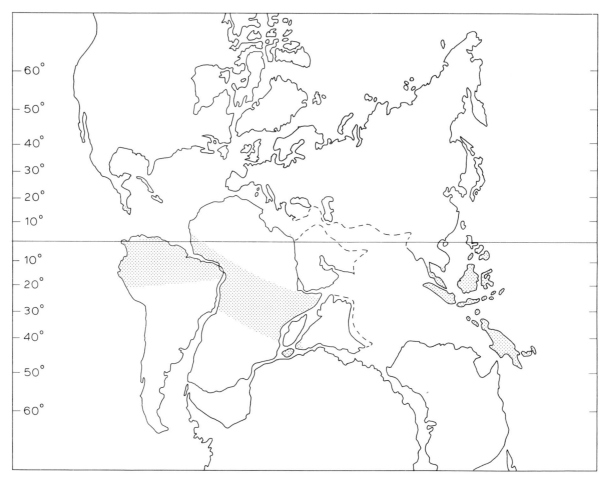

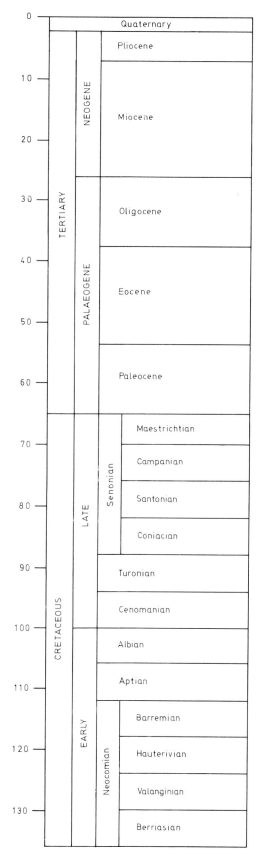

Figure 2.3 Stratigraphic diagram of Cretaceous and Tertiary stages. (Modified, after Wolfe, 1975)

pre-Cretaceous origin for the angiosperms'. The macro-fossil evidence, on re-examination, does not show that sudden increase in diversity during the Cretaceous which was previously thought to happen. Rather, there is a gradual accumulation of new taxa. The earliest definite fossil angiosperm pollen is *Clavatipollenites hughesii* from the Barremian of England (*see Figure 2.3* for stratigraphic diagram of Cretaceous and Tertiary stages). The pollen record also demonstrates a progressive increase in diversity during the Cretaceous (*Figure 2.4*, Muller, 1970), although the Jurassic records shown are regarded by Doyle *et al.* (1975) as from gymnosperms.

Further evidence regarding the first angiosperms comes from comparative morphology and anatomy of present-day species. Many attempts have been made

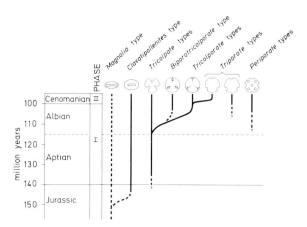

Figure 2.4 Primary radiation of fossil angiosperm pollen types. (After Muller, 1970)

to construct hypothetical evolutionary series, particularly relating to flower structure. Several of these regard the flower with unfused sepals, petals, stamens and carpels (as found in Ranales and Magnoliales) as primitive, and derive fused structures from these. The flowers of the catkin-bearing Amentiferae, in which calyx and corolla are generally more or less lacking, are regarded as reduced. Others, however, regard the Amentiferous type of flower as primitive, or even as requiring a separate phylogeny. The difficulty with series is that they can be read in either direction, or outwards from the middle in two directions.

Another idea is that teratological forms provide insight into evolution. Thus occasional flowers appear with leaves replacing petals, and these are held to be 'throw-backs', indicating evolution of petals from leaves. Such monstrosities are now beginning to be explained in physiological terms, and as evidence for evolution they are highly suspect.

In recent years the method of correlations has achieved some success as a means of indicating primitive angiosperm characters. Morphological characters are not randomly distributed through angiosperm families. Instead they frequently show marked correlation in their occurrence. One such group of characters contains many found in early fossil angiosperms, and is therefore regarded as indicating primitiveness

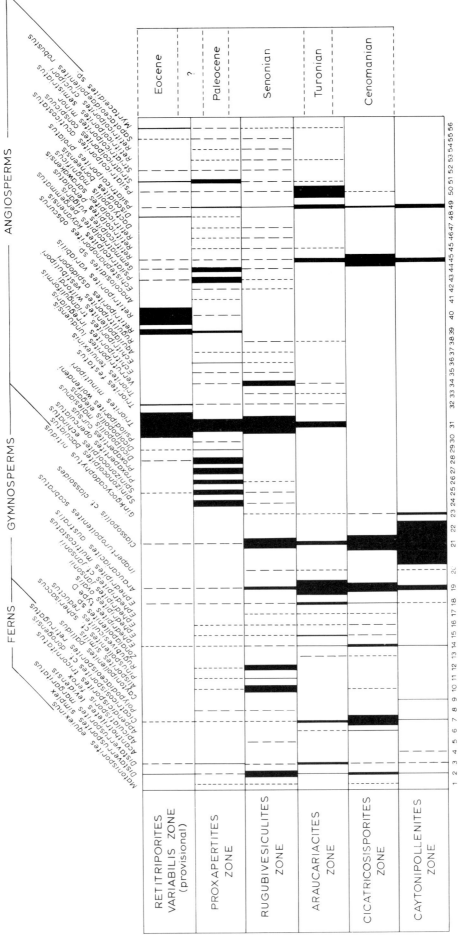

Figure 2.5 Ranges of pollen and spores in the Cenomanian to Eocene section of western Sarawak (Borneo) showing replacement of gymnosperms by angiosperms. (After Muller, 1968)

(Sporne, 1971). Possession or lack of these characters can be used to allot to each family a numerical 'advancement index' (Sporne, 1969). The Magnoliaceae, Austrobaileyaceae, Rhizophoraceae and Winteraceae come out with the lowest indices and are therefore believed to be the most primitive.

It is particularly interesting in relation to our subject that the families of dicotyledons inhabiting tropical rain forests have a low average advancement index (Wood, 1970). The figure obtained for tropical rain forests in Guyana was 45.1, for moist forest in Uganda 47.2 and for peat-swamp forest in Sarawak 47.5. These may not appear very different from the average of 56.6 for the world flora, but in fact the differences are highly significant with a probability of significance rising to 20 000:1 in the case of the Sarawak peat-swamp forest. This low advancement index leads one to expect that many of Sporne's 'magnolioid' characters would occur among the families of the rain forest in greater abundance than among the families of the world flora, and this is indeed true. Twelve 'magnolioid' characters which show highly significant positive correlation with occurrence in tropical rain forest have now been identified by Sporne (1973). In addition several of these characters show strong correlation with early occurrence in the fossil record, using both Muller's (1970) list of families whose pollen occurs before the Oligocene, and Chesters *et al.*'s (1967) list of families of which there are macrofossils in the pre-Tertiary. This evidence suggests that rain forest families are on average, rather primitive, although the alternative interpretation (Wood, 1970) that the 'primitive' characters simply reflect adaptation to the woody habit (which could be secondary) must not be forgotten.

Sporne's conclusion may seem to pre-empt discussion of the question of the *place* of origin of the angiosperms; obviously you might think this must have been the tropics. There is, however, no reason at all why the place of *survival* of the primitive angiosperms should be the same as their place of *origin*. The oldest definitely angiosperm pollen record is from England. The earliest macrofossil record of an angiosperm is *Celastrophyllum circinerve* from the Upper Barremian (Lower Cretaceous) of the U.S.A. (Chesters *et al.*, 1967). This early identification by Fontaine in 1889 must be treated with reserve, although the leaf may well represent an angiosperm of some kind. There is therefore no positive support as yet in either the macrofossil or the pollen record for the idea that angiosperms originated in the tropics. Of course, the actual place of origin of the angiosperm group may also be quite different from the place where it first came to dominate the vegetation. This latter process has often been regarded as strange and sudden. In fact, the replacement of a gymnosperm-dominated flora by one in which angiosperms are the main constituents was indeed sometimes rather rapid at one place, but occurred at different times within the Cretaceous in different places. From Albian to Senonian times the angiosperms steadily and progressively occupied a major part of world vegetation. The whole process

took perhaps 25 M years, which is neither fast nor slow (Hughes, 1961). The direction of this march of the angiosperms has been investigated by Axelrod (1959) and Smiley (1967). From about 30°N to 70°N in North America and western Europe there are floras showing progressively later arrival of angiosperms with increase in latitude. Unfortunately the data from other areas and especially from within 30° of the Equator are too scanty to permit any firm conclusion.

One good example from the equatorial region of the replacement of gymnosperms by angiosperms is that from the Pedawan and Plateau Sandstone formation in Sarawak, Malaysia, studied by Muller (1968). The floral (pollen) succession described covers approximately 40 M years, from the Cenomanian (Upper Cretaceous) to the Eocene. The section cannot therefore be complete, and the period not covered because of stratigraphic gaps may exceed the period represented. Nevertheless, the results, shown in *Figure 2.5*, present a coherent story of the emergence of an angiosperm flora in an equatorial region. Gymnosperms, represented particularly by *Classopollis* cf. *classoides* and *Araucariacites australis*, are replaced gradually, but particularly during the Cenomanian and Senonian, by *Triorites minutipori* (which has possible affinities with Juglandaceae and Ulmaceae) and over 30 other angiosperm types, including some showing possible affinities with Palmae (*Nypa, Calamus*), Apocynaceae, Ulmaceae, Santalaceae, Olacaceae, Aquifoliaceae, Anacardiaceae, Tiliaceae, Sapotaceae and Myrtaceae. This replacement appears to have been a competitive one; there is no evidence that the gymnosperms died out, leaving a vegetational vacuum.

The date of this change – Cenomanian to Senonian or thereabouts – is rather later than would be predicted by Axelrod's (1959) theory that the angiosperms dominated the tropics first and thereafter spread poleward, but such anomalies are not unlikely with the present paucity of data. A single succession is unlikely to be representative of any broad latitudinal belt, and there is no over-riding reason why all parts of a latitudinal belt should have been invaded by angiosperms simultaneously. The balance of the evidence probably still suggests that very early angiosperm-dominated floras may eventually be discovered in tropical regions.

The flora from Borneo should certainly not be taken as necessarily typical of equatorial regions. This is particularly so since South-East Asia was distant from Africa and South America in the Late Cretaceous, and it is therefore necessary to consider evidence from other areas. This is once again chiefly palynological since macrofossils are rare and notoriously difficult to determine. We may note in passing, however, the macrofossil evidence for occurrence of probable monocotyledons in both Africa and South America in Late Cretaceous time. From Africa we have *Typhacites latissimus* from the Enugu formation, in Nigeria, of Upper Cretaceous or Lower Tertiary age (Seward, 1924), while from Colombia comes a record of a possible *Musa* sp. in Cretaceous rocks (Huertos

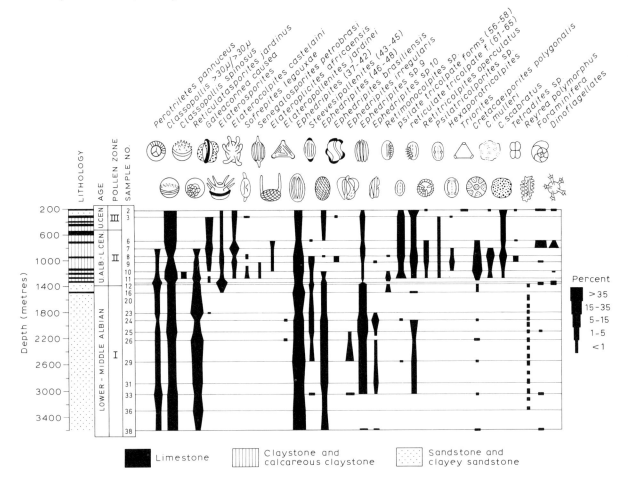

Figure 2.6 Pollen diagram from the Cretaceous of Brazil showing replacement of gymnosperms by angiosperms. Only selected taxa are shown. (After Herngreen, 1973)

and van der Hammen, 1953). *Musa* is not generally considered to be a native genus in the Americas, so the record must be regarded as surprising and perhaps doubtful; but it presumably represents a monocotyledon of some sort.

The Cretaceous fossil microfloras of Africa and South America show a marked similarity. In the middle Cretaceous of Portuguese Guinea and Senegal there is abundant gymnosperm pollen, particularly *Ephedripites* and *Steevesipollenites*. Some of the *Ephedripites* grains approach very closely the modern *Ephedra* pollen (Stover, 1964). Similar results are reported from the Ivory Coast and Senegal (Jardiné and Magloire, 1965) and from Gabon (Jardiné, 1974). From Brazil, Herngreen (1973, 1975) reports an abundance of *Ephedripites* in rocks of Albian to Cenomanian age (*Figure 2.6*). The widespread occurrence of plants related to *Ephedra* and the very limited occurrence of Pteridophyte spores might suggest arid conditions. Many of the taxa from the Brazilian samples were also reported in Africa. A similarity was also noted by Brenner (1968) between samples from Peru and those from Africa. The general similarity of these floras is not surprising in view of the likely close proximity of Africa and South America at the time (*Figure 2.2*). The similarity continues into Senonian times when *Buttinia*, for example, occurs in both regions

(Germeraad *et al.*, 1968). The Indo-Malesian flora was apparently different (Muller, 1968), which is not surprising in view of its geographical isolation from the other equatorial regions at the time.

In Borneo the big jump from 16% to 47% angiosperms happens around the Cenomanian–Turonian boundary. Herngreen's (1973, 1975) data from Brazil show a marked increase in angiosperms in Albian–Cenomanian time (*Figure 2.6*) and a second increase between the Cenomanian and Senonian. Similar results were obtained from Brazil by Regali *et al.* (1974). African floras probably paralleled South American ones (Boltenhagen, 1976; Petrosyants and Trofimov, 1971) so the take-over by the angiosperms can be located as approximately Albian to Senonian in all three equatorial regions.

The mention above of the possible botanical affinities of certain pollen types must not be taken to indicate that Late Cretaceous vegetation was necessarily at all modern in appearance. In the first place these botanical affinities are totally unproven and are likely, except in the uppermost Cretaceous, to remain so. Secondly there are no pollen types identical with those of extant taxa, and there is a large number of pollen types in the record which do not resemble any modern types at all. There was a great proliferation of some of these extinct types in the Senonian. For instance,

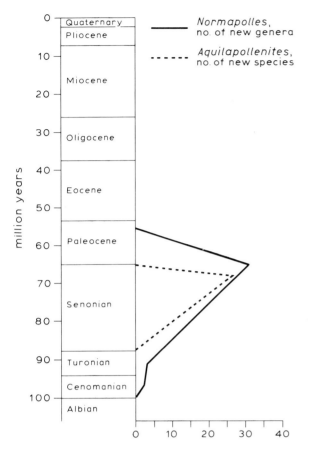

Figure 2.7 Radiation of two extinct angiosperm pollen groups. (After Muller, 1970)

Figure 2.8 The positions of the continents in Eocene time about 50±5 M years B.P. The shaded areas are those now within 10° of the equator. Palaeolongitude is not indicated because there is no way of measuring it. Mercator projection. (After Smith et al., 1973)

over the world as a whole, over 30 new *genera* of fossil pollen types in the 'Normapolles' group appeared in the geological record during the Senonian (*Figure 2.7*). *Aquilapollenites* shows the same features on a lesser scale.

2.3 THE PALAEOGENE
(Paleocene, Eocene and Oligocene Epochs)

The three epochs included here lasted altogether about 40M years, i.e. from 65—26 M years ago (Harland *et al.*, 1964). During such a long period of time there was considerable movement of the plates that make up the earth's crust, so it is impossible to show their positions adequately in a single map. For a first approximation, however, we may consider the positions in the Eocene (*Figure 2.8*). The Atlantic was by this time a wide rift, Africa had almost reached its present position, and the Isthmus of Panama was almost, but not quite, continuous. Stepping stone islands were probably present there (Maldonado-Koerdell, 1964). The Andes were low or unformed. India was still in the middle of the Indian Ocean and the Himalayas not yet elevated, according to Smith, A.G. *et al.*, (1973). Powell and Conaghan (1973), however, present evidence that India had already collided with Asia by the Middle Eocene. The Sunda platform of South-East Asia was in its present position, although largely submerged, but Australia and its northern continuation, New Guinea (the latter also submerged at the time) were still far to the south.

Although a large number of macrofossils has been reported from the equatorial Palaeogene, their value as evidence is not proportionately great. Some of the early publications describe new species from a fragment of a single leaf. In many cases determination is based only on macro-observation of the specimen,

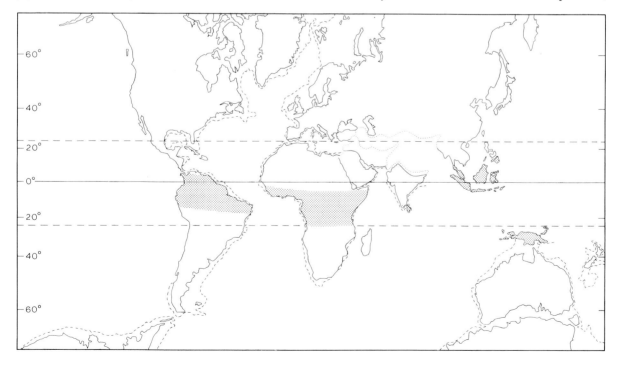

Table 2.1 COMPARISON BETWEEN THE LONDON CLAY FLORA AND MODERN TROPICAL FLORAS (AFTER REID AND CHANDLER, 1933; PEARSON, 1964)

	Exclusively tropical (%)	*Mainly tropical (%)*	*Equally tropical and non-tropical (%)*	*Mainly extra-tropical (%)*
Living tropical families	15	32.5	32.5	20
London Clay families	11	32	46	11

without recourse to cuticular preparations or sectioning. In addition, many of the descriptions are of isolated finds which are of little value in the attempt to reconstruct the flora and vegetation. This situation presents something of a contrast with that in north temperate regions where several Palaeogene macrofloras have been examined. Examples are the London Clay flora of Britain (Eocene; Reid and Chandler, 1933), and, from the United States, the Goshen flora (Eocene–Oligocene; Chaney and Sanborn, 1933), the Comstock flora (Upper Eocene; Sanborn, 1935), the La Porte flora (Oligocene; Potbury, 1935), the Florissant flora (Oligocene; MacGinitie, 1953), and the Weaverville flora (Lower Oligocene; MacGinitie, 1937).

Although all these are from well outside the equatorial regions they merit brief mention because the floras frequently show marked tropical affinities. For instance the London Clay flora includes the characteristic fruits of the palm *Nypa*, at present confined to South-East Asia, along with many other taxa of Indo-Malayan affinity. A comparison of the whole flora with the present world tropical flora shows marked similarity (*Table 2.1*). The affinities of the Goshen macrofossils are partially with the flora of neo-tropical regions. This can be demonstrated to some extent by the determinations, but the nature of the flora was also suggested by statistical analysis of the leaf characters, compared with those from a modern Panamanian forest flora, albeit of only 41 species. The results are shown in *Figure 2.9*. Other Eocene floras have led Dilcher (1973) to conclude that an equable warm temperate to cool sub-tropical environment is indicated, rather than a fully tropical one.

The Oligocene floras are more equivocal and some at least (e.g. the Florissant flora) seem to be of only temperate affinities (Chaney, 1947). Wolfe and Hopkins (1967) regard these as indicating climatic fluctuations. Axelrod and Bailey (1969), however, claim that the different source areas of the fossils preserved (upland, lowland, swamp etc.) could be responsible for these differences. It seems clear, nevertheless, that early in the Palaeogene, present day tropical floras were at least strongly represented in temperate regions, and even as far north as what is now 50°N (Chaney, 1947).

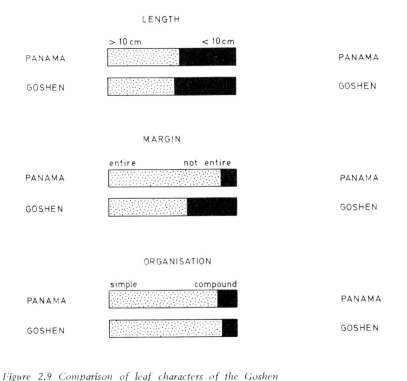

Figure 2.9 Comparison of leaf characters of the Goshen Eocene–Oligocene flora with those of the modern Panamanian flora. In most respects the two are very similar indeed. (Data from Pearson, 1964)

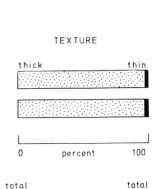

It is all very well to find tropical floras in temperate regions, but what was present in the tropical regions at this time? For South-East Asia most of the Tertiary fossil macroflora was listed by Posthumus (1931). In view of the uncertainties of identification and date of many of the fossils, practically all that can be concluded from this list is that Palmae, Fagaceae, Moraceae, Leguminosae and Dipterocarpaceae were probably common in the Tertiary floras. There is a fossil *Nypa* fruit from central Borneo, of possible Eocene age (Kräusel, 1923).

The Deccan Intertrappean beds of India, usually thought to be mostly of Eocene age, contain a rich flora of wood and leaves including many which have been referred to modern families and genera. The families identified include Podocarpaceae, Araucariaceae, Palmae (including *Nypa*), Musaceae, Euphorbiaceae, Sonneratiaceae, Tiliaceae, Simaroubaceae, Burseraceae and Lecythidaceae (Lakhanpal, 1970).

From equatorial Africa there are few macrofossils. Koeniguer (1971) described three fossil timbers of Paleocene age from Niger, including representatives of Euphorbiaceae and Meliaceae. Fossil wood has also been found in Kenya (Bancroft, 1933a) but was not determined in detail.

There are several fossil macroflorulae from the equatorial Americas. Fruits or seeds referred to *Anacardium*, *Carpolithus*, *Celtis* and *Sapindoides* are recorded from the Eocene of northern Colombia (Berry, 1942). Of late Eocene to Oligocene age are fruits from western Ecuador: *Anacardium*, *Annona*, *Astrocaryum*, *Palmocarpon*, *Sapindoides* and *Vantanea* (Berry, 1929). Upper Eocene beds in Venezuela (Berry, 1939) yield fossils remarkably similar to those from the same period in the southeastern United States, about 25° further north. Seventeen species are recorded from various beds in the Eocene to Miocene of the Canal Zone (Berry, 1919). The generic attributions are less useful than the fact that, of the thirteen species represented by leaves and leaflets, all have entire margins. On the other hand, of the four leaves with unbroken tips, none has a pronounced drip-tip. The sample is too small to be of statistical significance, but is perhaps not inconsistent with a flora broadly similar to the present one. Fruits of *Nypa* have been recorded from the Paleocene of Brazil (Dolianiti, 1955), and from the Eocene of Texas (Tralau, 1964).

The general impression given by all these early Tertiary macrofloras is very vague indeed. Certainly the floras were becoming gradually more modern, because by contrast with Cretaceous material, it is possible to recognise distinct affinities with present-day taxa, and even to put fossils into the same genera as modern species (e.g. *Anacardium*). Any close comparison with modern floras is, however, difficult to justify. The most that can be said about these equatorial macrofloras is that there is nothing about them that *demands* radical differences of climate or geography between then and the present time. It is however perfectly consistent with these fossils that Paleocene equatorial climates were somewhat cooler, or considerably warmer, than at present.

The microfossil evidence from the equatorial Tertiary has been ably presented by Germeraad *et al.* (1968), and the botanical significance of this reviewed by Muller (1970). There is therefore little sense in considering the general evidence separately for the Palaeogene and Neogene, which will accordingly be dealt with together in the final section of this chapter.

There are, however, one or two points more specific to the Palaeogene, which may be dealt with here. The first of these is the conflict of evidence regarding the presence of *Nypa* and other taxa of Indo-Malesian affinity in the London Clay. *Nypa* is now recorded (Tralau, 1964; Germeraad *et al.*, 1968) from the Paleocene and/or Eocene of West Africa (Senegal), France, Britain, Italy, Poland, Southern U.S.S.R., Egypt, Central India, Borneo (pollen), Brazil and Texas. It is also known from the Miocene of Assam, and Couper records its pollen in the Pliocene of New Zealand although J. Muller (personal communication) rejects this record. There are also records from South Australia. Its present main distribution is Ceylon, Assam, Burma, Malesia (including New Guinea) and China. The numerous Eocene records for Europe seem to confirm the presence of *Nypa* on the north side of the Tethys, and this fits the evidence for a 'tropical' climate of the rest of the London Clay flora and of the tropical corals of the time (Tralau, 1964). On the other hand, van Steenis (1962b) has suggested that the London Clay flora might represent material drifted across the Tethys. The contribution of palynology to this discussion is important. European Eocene microfloras are very different from the few Indo-Malesian ones (Muller, 1970). Although it is dangerous to argue from absence, this evidence contrasts strikingly with the idea of tropical forest in southern Europe. At most, there is evidence of a few Indo-Malesian taxa penetrating Europe in warm phases (Krutzsch, 1967). Daley (1972) has suggested that the London Clay climate was one not represented anywhere today: seasonal but frostless, with rainfall higher (for the latitude, then 40°N) than today, and warmer than now but not as warm as in tropical rain forest areas.

Nypa is a monotypic genus of sub-aquatic palms which grow in brackish water, usually at the landward side of mangrove swamps. In South-East Asia the record of pollen from deposits formed offshore is now sufficiently complete to permit reconstruction of the history of this mangrove vegetation (Muller, 1964, 1968, 1972). Of the seven distinct mangrove pollen types in the fossil record of north-west Borneo (*Figure 2.10*) only two, *Nypa* and *Brownlowia*-type are present from the Paleocene. There then follows the progressive appearance of the *Rhizophora*-type in the Oligocene, the *Sonneratia caseolaris*-type and the *Avicennia*-type in the Miocene, and the *Sonneratia alba*-type in the Late Miocene/Pliocene. Thus the gradual build up of a modern plant community can be traced. Indeed the pollen suggests that actual evolution of some species was occurring *in situ*. The fossil pollen genus *Florschuetzia* shows a range of forms which change in abundance in a way consistent with evolutionary change, culminating in two pollen

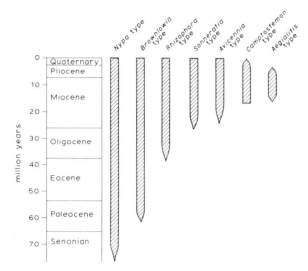

Figure 2.10 The distribution of mangrove pollen types through geological time in Borneo. (Modified, after Muller, 1972)

types which are recognisable as the present day *Sonneratia caseolaris*-type and *Sonneratia alba*-type (*Figure 2.11*).

Borneo has also been the site for another palynological discovery of biogeographical interest. The Oligocene and Miocene pollen assemblages, mostly derived from shallow-water deposits, include a proportion of montane pollen types (Muller, 1966). There is nothing unusual in this, though it is useful in indicating that there have been mountains in the Borneo region for a long time; but what is strange is that the types represented (*Pinus, Picea, Tsuga, Ephedra* and *Alnus*) are all mainland Asiatic types which do not occur in Borneo today. The Australasian-centred types which dominate the montane forests of Borneo today (represented in the pollen record by *Dacrycarpus imbricatus*-type and *Phyllocladus*-type)

do not appear in the record until the Plio–Pleistocene boundary, by which time the Asiatic types have disappeared. These findings are partly explained by continental movement. Reference to *Figure 2.2* will show the Australasian continent far to the south in the Palaeogene, while Borneo, in so far as it was above water at the time, was essentially part of the Sunda peninsula of Asia. The arrival of the Australasian types in the Bornean flora would have followed the drifting of Australia to approximately its present position in the Neogene.

The Palaeogene floras suggest that considerable similarities existed between all the tropical regions at the time. Not only *Nypa* but also *Proxapertites* and *Echitriporites* were pan-tropical in the Paleocene, and by the Oligocene several other taxa were of equally wide occurrence.

2.4 THE NEOGENE
(Miocene and Pliocene Epochs)

This interval lasted from 26 M years ago to about 2 M years ago (Harland *et al.*, 1964; West, 1977) and the continents can be taken to have reached approximately their present positions during it. This was the time during which the Isthmus of Panama was formed (Maldonado-Koerdell, 1964; Raven and Axelrod, 1974). It was also a period of great mountain-building (the Alpine orogeny) which continued on into the Quaternary. The Andes of South America originated in the Neogene or later (van der Hammen, 1974) and the Himalayas, if they had not already formed, probably also uplifted at this time as the Indian plate collided with Asia. Some at least of the East African volcanoes are no older than Neogene and Mount Kinabalu of Borneo gives radiometric dates mostly in the range 9–1.5 M years (Jacobson, 1970). Many New

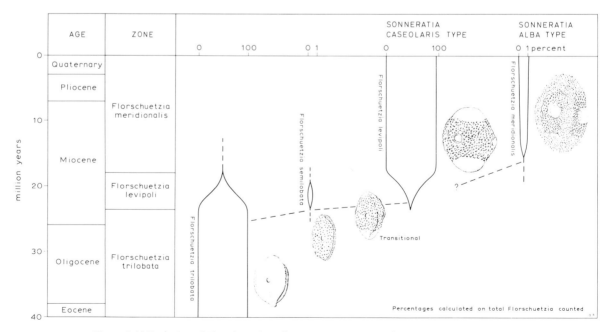

Figure 2.11 Evolution of Florschuetzia pollen types in Borneo. (After Germeraad, et al., 1968)

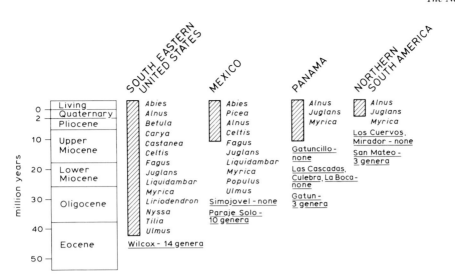

Figure 2.12 The occurrence of pollen of temperate trees in Tertiary deposits of Central America showing the gradual migration southwards. (After Graham, 1973)

Guinea mountains are probably even younger (i.e. Quaternary) as are the Javanese volcanoes (van Bemmelen, 1949).

Satisfactory accounts of macrofossil floras of Neogene age, as opposed to isolated fossil finds, are not numerous from equatorial regions. One of the better ones is Kräusel's (1929) study of plants from South Sumatra, in which several cuticle preparations are figured. Although the individual determinations may be doubted, it is noteworthy that of the 13 different dicotyledonous leaves figured with the margins showing, all have entire margins. This, albeit inadequately, supports Kräusel's idea that the flora is essentially of modern rain forest type. The well-known plant fossils from the Batu Arang Neogene coal of the Malay peninsula were studied by Ridley (reported in Stauffer, 1973) who determined eight taxa resembling those of present day South-East Asian rain forest. The preponderance of leaves of coriaceous type could well be the result of selective preservation.

From equatorial Africa there are a few Neogene fossil timbers (Bancroft, 1932, 1933a, b), the Miocene flora of Rusinga Island, Lake Victoria (Chesters, 1957), a flora of 12 species from the Bugishu sandstones, Uganda (Chaney, 1933), and a small flora from the volcanic deposits of Bukwa, Uganda (Hamilton, 1968). There are also some fossil timbers from the Sahara, outside the equatorial area (Louvet, 1973). The most interesting determination, from a biogeographical standpoint, is Bancroft's (1933b) claim of affinity of some material to the Dipterocarpaceae, group Dipterocarpoideae, which are absent from Africa today.

From the Neogene in equatorial South America many macrofossils have been described. Unfortunately most of these occur as individual finds rather than floras. Berry (1936), however, lists 12 species of Miocene age from one site in Colombia. All are of present-day rain forest and mangrove type, and many of them have also been recorded elsewhere in the Miocene of South America. The island of Trinidad has yielded several Neogene macrofossil floras (Berry,

1925, 1937a, b). The determinations must be treated with reserve, but the leaves illustrated are almost all with entire margins, which is consistent with a rain forest flora.

The Cuddalore sandstone of Southern India is of doubtful age: possibly Mio-Pliocene, but possibly as early as Eocene. It contains a fossil wood flora from which have been determined members of the Dipterocarpaceae, Leguminosae, Combretaceae, Euphorbiaceae, Podocarpaceae, and Palmae (Prakash, 1965). This suggests a flora of modern type existing in a moist tropical climate — possibly more moist than in the area at present (Prakash, 1973).

The general impression gained from all these Neogene macrofloras is of an equatorial region largely dominated by rain forest. This conclusion is probably unjustifiable, however, because there is usually less chance of plants in arid environments entering the waterlogged environments that are necessary for their preservation as fossils.

The bulk of the microfossil evidence will be considered in the next section together with that from the Palaeogene, but there are a few points which refer specifically to the Neogene and may be mentioned here.

The construction of the Isthmus of Panama in Neogene times had far-reaching biogeographical results. The very different faunas and floras of North and South America were at last in direct contact. Floral migrations on a vast scale were probably inhibited by the very different ecological conditions to which the floras were adapted: the South American humid warm environment contrasting sharply with the Mexican desert and the North American temperate environments. At the same time, however, the Andes were rapidly being uplifted, affording a corridor for the movement of montane taxa of temperate affinity.

The macro- and microfossil evidence (Graham, 1973) from various points on the Isthmus now allows us to trace some of these movements (*Figure 2.12*). *Myrica, Juglans* and *Alnus* had migrated southwards into the Isthmus by the Pliocene and had reached

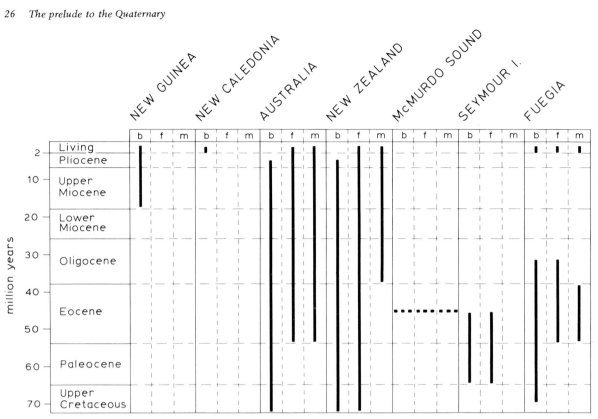

Figure 2.13 The recent (i.e. present) occurrence and fossil record of Nothofagus. *b,* brassii *type; f,* fusca *type; m,* menziesii *type. The fossils from McMurdo Sound are unspecified as to type. (After van Steenis, 1971)*

South America by the Pleistocene. *Quercus* (not shown in *Figure 2.12*) followed them later. Other temperate genera, *Liriodendron, Castanea* and *Tilia*, for some reason did not succeed in expanding in this way. It is possible that the expansion of *Podocarpus*, essentially a southern genus, into North America occurred during the same period, although its ancient origin means that it could have made the journey very much earlier.

Mountain building was also in progress on the other side of the Pacific, in New Guinea, as Australasia continued its northward drift. *Nothofagus* is abundant in the mountain forests of New Guinea (Hynes, 1974). The species of this 'key genus of plant geography' (van Steenis, 1971) fall into three groups whose pollen is distinguishable (Cranwell, 1963). The present and former distributions of the three groups are summarised in *Figure 2.13*. There are morphological reasons for regarding the *brassii* group as the most primitive (van Steenis, 1971) and its wide distribution relatively early in the fossil record supports this. In New Guinea only the *brassii* type occurs at present, and then only in the mountains. Its pollen was certainly abundant in New Guinea in the Miocene (Khan, 1974), along with that of the southern conifer *Dacrycarpus*. The latter managed to migrate to other parts of South-East Asia but *Nothofagus* apparently did not expand westward beyond New Guinea. Actually, Khan's samples include a few grains of *Nothofagus fusca*-type pollen. As, however, the genus has record-breaking propensities for long-distance pollen dispersal, for example 4000 km from South America to Tristan da

Cunha (Hafsten, 1960), these grains may be regarded as wind blown. In New Guinea *Nothofagus* spp. are at present restricted to altitudes above 600 m (Hynes, 1974). It may well be that their survival in New Guinea depended on the formation of montane refuges as the drift towards the equator made the New Guinea lowlands too hot for them, or in places too dry for them.

There have been several records of *Nothofagus* pollen in the northern hemisphere (e.g. Zaklinskaya, 1964; Elsik, 1974), but these are best disregarded. Some may possibly be misidentifications (Petrov and Drazheva-Stomatova, 1972), and others might result from long distance transport.

2.5   PALYNOLOGICAL EVIDENCE AND SYNTHESIS

Although the overall picture of equatorial vegetation in the last 120 M years is very incomplete, some tentative generalisations may be offered, if only perhaps in the hope of stimulating further research. The place and time of origin of angiosperms and the precise timing of their later take-over of the world's vegetation are still not certainly known. Preliminary evidence is inadequate to say whether the equatorial regions played a prime role in either of these events. Of the later differentiation of angiosperms into their many modern taxa, we can now say a little, thanks chiefly to the work of palynologists. Pollen grains recognisable as closely similar to those of living angiosperms first appear (on a world basis) in the very late Cretaceous,

Figure 2.14 Range chart of identified pollen types showing the gradual appearance of the modern angiosperm flora. (After Muller, 1970)

and include types referable to *Nypa, Nothofagus* and *Ilex*. The progressive appearance of other types is summarised in *Figure 2.14*. One of the interesting features of the record, which does not conflict substantially with the macrofossil record, is the gradual, as opposed to sudden, appearance of new types. There are certainly times when many new types appear, such as the Upper Miocene, but there is a steady sprinkling of new records at all times. It is of particular interest that the fossil history of many 'advanced' families is so short. For instance, the Compositae, with their distinctive pollen, are not known before the Lower Miocene, when the subfamily Tubuliflorae made its appearance somewhat before the Liguliflorae. Although the family may have existed before that time it was almost certainly not widespread until the Miocene. Thus our present equatorial floras did not arise suddenly; they are the resultant of the gradual

blending together of many taxa as these evolved and migrated, as well as of the loss of other taxa by extinction. Hence the suggestion which is sometimes made that our present tropical rain forests are of great geological antiquity can be only partially true. The vegetation of equatorial regions has probably been in a dynamic state for a very long time indeed.

There have been many attempts to use vegetational evidence to reconstruct the world Tertiary climates, and I do not intend to make another one here. In fact equatorial evidence has not been widely used in such reconstructions, since it is still rather fragmentary. An exception to this is the use of ocean cores, either for $^{16}O/^{18}O$ palaeoglaciation measurements (Emiliani, 1971; Shackleton, 1967), or for their fossil content, especially foraminifera (e.g. Bandy, 1968). Clearly, however, the full reconstruction of Tertiary climates must await further research results.

3
The Quaternary Vegetation of Equatorial Africa

3.1 INTRODUCTION

The meagre fossil evidence brought forward in the previous chapter does little to advance our knowledge of the state of equatorial African vegetation at the start of the Quaternary. The few fossils available are mostly of rain forest types, and these are found outside the present rain forest area, which might suggest a former greater extent of rain forest. The records for *Dipterocarpoxylon* (Bancroft, 1933b) and *Nypa* (Tralau, 1964), representing taxa extinct from Africa today, might also suggest a richer flora, one perhaps resembling more closely the South American and Indo-Malesian floras. This is in great contrast with the situation today.

3.2 PRESENT VEGETATION

From the distribution map shown in *Figure 1.5* it will be noted that the rain forest covers a relatively small area in Africa, and this is in two parts. Both parts receive their annual rainfall of more than 1500 mm mainly from the humid maritime air brought in by the south-westerly winds. The zone of forest–savannah mosaic between (the 'Dahomey Gap') is explained by Aubréville (1949) as being related to the fact that the coast itself runs in a WSW–ENE direction at that point, so that the wind makes an acute angle with the coast and little moisture is brought ashore. The eastern part of the area also receives some rainfall from the east coast winds.

The areas north and south of the rain forest bear vegetation which is likewise largely climatically controlled. These areas are under the influence of either the very dry Saharan or the less dry South African anticyclone. The result, on both sides of the equator, is a gradual change from rain forest to forest–savannah mosaic, open woodland, dry savannah and grasslands, and finally semi-desert and desert. These changes are correlated with a reduction in total rainfall as well as with more pronounced dry seasons. Similar vegetation occurs on the east side of Africa in the equatorial zone.

There are two main high mountain areas in equatorial Africa. In West Africa lies the isolated Cameroon Mountain, rising to 4070 m. This elevated area bears a dwarf, montane type of forest (Keay, 1955). The montane flora is related to the East African montane flora and also to the Eurasian flora (Morton, 1962). This disjunction between East and West African montane floras is of great biogeographical interest.

The East African high mountains consist of a series of isolated, chiefly volcanic peaks on the East African plateau, roughly surrounding Lake Victoria, and the separate high plateau of Ethiopia. These eastern mountains rise to 5895 m (Kilimanjaro) and bear permanent snow. Since the East African plateau is the area where most African palynological research has taken place, the vegetation of this area is of considerable interest.

As a mountaineer ascends an East African mountain he finds the lower slopes, in so far as they have not been cleared for grazing or cultivation, vegetated by a forest which is either fully evergreen, in moist areas, or semi-deciduous, in drier areas. As he climbs, the forest around him becomes gradually shorter and less diverse, so that about 1500–2000 m distinct dominants begin to appear. *Popocarpus milanjianus* is one of the chief species here in most areas, although it is, for some unknown reason, absent from wet sites in Uganda. It is accompanied in many dry areas by trees of *Juniperus procera*, with which it may be co-dominant. In some wet areas *Pygeum (Prunus) africanum* is the dominant tree. The mountaineer may well come across a fairly distinct bamboo zone, consisting largely of enormous groves of *Arundinaria alpina* which grows to 15 m or more and is much used by man. At higher elevations, around 2800 m, the forest is dwarf – often more of a thicket than a forest. Prominent here are the tree species of the genus *Hypericum* (represented by herbs in temperate regions), the small rosaceous tree *Hagenia abyssinica* and a number of shrubs or small trees of the genus *Rapanea*. At approximately 3000 m there is another gradual change to a shrubbery dominated by *Philippia trimera*, one of the 'tree heathers', *Helichrysum* spp. and several other shrubs. On some mountains there is an abundance of *Stoebe kilimandscharica*, a shrub with an ericoid habit but belonging to the Compositae. There are also species of the fantastic sub-genus *Dendrosenecio*. This is a section of *Senecio*, the ragworts of temperate regions, but is pachycaulous or palm-like in its growth form. Limited branching permits individuals to assume the shapes of bushes up to c. 8 m high. Extreme pachycauly, with virtually no branching, is exhibited also by several species of *Lobelia* in this vegetation. The possible significance of pachycauly in this environment has been discussed by Cotton (1944), Hedberg (1964) and Coe (1967), who regard it as an adaptation to the wide diurnal temperature variation. The large leaf rosette tends to close at night in both *Dendrosenecio* and *Lobelia*, and Hedberg found that on a typical frosty morning when the outside temperature was $-4\,^{\circ}$C, the inner part of the rosette near the growing point had remained at $+1.5\,^{\circ}$C. Pachycauly may thus be a protection against lethal night temperatures. This may be the physiological result of pachycauly, but that is not necessarily the same as the evolutionary cause. The giant *Lobelia* spp. probably evolved from species of temperate affinity, and the fact that they are usually monocarpic, i.e. flower once and then die, suggests their ancestors could have been biennials. Most biennials are stimulated to flower either by day length changes or by the cold shock of the temperate winter. When kept under artificial conditions which prevent them receiving these stimuli, they continue to grow and assume monstrous forms, often pachycaulous. When eventually stimulated, they flower and die. It therefore seems possible that the pachycaul habit arose on tropical mountains because the seasonal changes were inadequate to stimulate flowering, and that the habit subsequently became genetically fixed. An alternative theory is proposed by Mabberley (1975).

The giant ragworts and lobelias persist well up above the limit of real shrubbery among a vegetation with abundant grasses, especially the tussock-forming *Festuca pilgeri*, and many other herbaceous species. This is the Afro-alpine grassland of Hedberg (1951) and it persists almost to the thermal limit of plant growth at about 5000 m altitude. A few isolated plants are found even higher, and Greenway (1965) records *Helichrysum newii* at 5760 m on Mt Kilimanjaro.

Hedberg (1951, 1964) has made extensive studies of the individual mountains and has classified the vegetation into three altitudinal belts, the Montane Forest Belt, the Ericaceous Belt and the Afro-alpine Belt. Within the belts, distinct zones are recognised, thus the Montane Forest Belt contains, in ascending order, the Montane Rain Forest Zone, the Bamboo Zone and the *Hagenia-Hypericum* Zone; the Ericaceous Belt contains the Moorland Zone and the Ericaceous Zone; the Alpine Belt is not divided. Each Belt is represented on every mountain (provided it is of sufficient height), but not all the individual Zones are necessarily present.

In addition, there are striking variations in the altitudes of the Zone boundaries on different mountains and even on different parts of the same mountain. Some of these differences may be related to moisture availability. For instance, on Mt Meru the Bamboo Zone is prominent on the wet southern slopes of the mountain but absent on the drier northern side. Some of this variability is illustrated in *Figure 3.1*. This diagram does not adequately draw attention to the variability in the Montane Rain Forest Zone. In Uganda, it is possible to distinguish at about 2000 m altitude a moist montane forest, dominated by *Pygeum (Prunus)* sp., from a dry montane forest chiefly of *Juniperus procera* and *Podocarpus gracilior*. The general relationships of these and other types of montane forest are illustrated in *Figure 3.2*. In this diagram the High Montane Heath and High Montane Moorland correspond approximately to the Ericaceous Belt and Afro-alpine Belt of Hedberg. The 'Moist Montane Forest B2' corresponds to the *Hagenia-Hypericum* Zone.

No zonation scheme is completely adequate to describe the vegetation, and for the interpretation of pollen diagrams it may be important to know the present altitudinal ranges of various taxa as well as the overall vegetation type. Fortunately this information has been collected for highland Uganda by Hamilton (1972), whose results are shown in *Figure 3.3*. Of course these overall ranges hide important ecological differences. The importance of aspect, for instance, is demonstrated by the finding that altitudinal ranges differ considerably on wet and dry sides of, for example, Mt Elgon (*Figure 3.4*). It appears that under dry conditions (the eastern slopes) species reach higher altitudes than on the wetter western slopes.

It is also important to realise that the African mountains have been subject to disturbance by man over a very long period. Burning and grazing at all altitudes have resulted in replacement of woody by herbaceous vegetation, and the whole of Hedberg's 'Moorland Zone' may well be secondary. Much of the tussock grassland of the Afro-alpine Belt is also

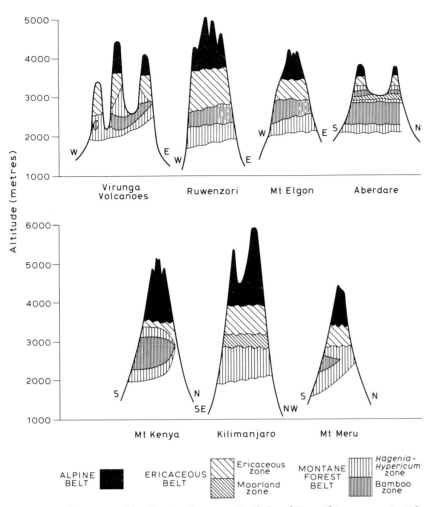

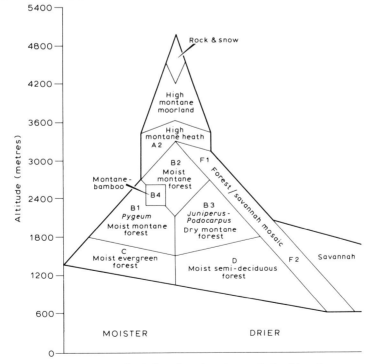

Figure 3.1 Schematic profiles showing the vegetation belts of East African mountains. The wettest side of each mountain is turned to the left, and compass directions are indicated at the base of each profile. (After Hedberg, 1951)

Figure 3.2 Generalised altitude and moisture relationships of the montane forest vegetation of Uganda. (After Lind and Morrison, 1974)

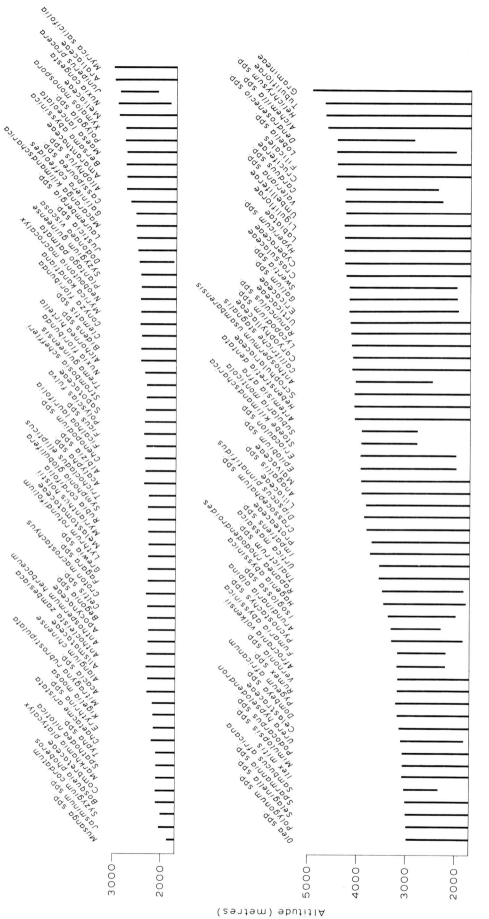

Figure 3.3 Altitudinal ranges of some plant taxa in highland Uganda. (After Hamilton, 1972)

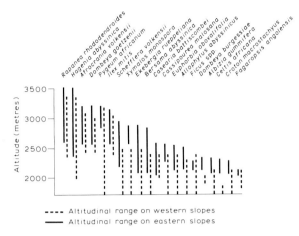

Figure 3.4 Altitudinal ranges on Mt Elgon of the commoner trees and shrubs which occur on both eastern and western slopes. (After Hamilton, 1972)

subject to fires, but it is possible some of these are natural (Greenway, 1965). Disturbance is also a feature of all the lowland vegetation types to a greater or lesser extent. Often this has the effect merely of sharpening the boundaries of climatically controlled vegetation types, but in extreme cases it may be responsible at least for the maintenance and perhaps even for the origin of the vegetation.

3.3 BIOGEOGRAPHICAL PROBLEMS

The most interesting biogeographical problems of this vegetation may be listed as follows:

1. The depauperate nature of the rain forest flora compared with those of South America and Indo-Malesia (Richards, 1973).
2. West African rain forest — East African disjunctions (Hamilton, 1974):
   (a) A few isolated patches of forest on the East coast have species whose next occurrence is in West Africa e.g. *Adiantum confine, Paramaclobium caeruleum*
   (b) Numerous West African species have outliers in western Kenya or Tanzania e.g. *Bolbitis gemmifera, Chrysophyllum perpulchrum*
   (c) Two species occur as different sub-species in West Africa and East Africa — *Greenwayodendron suaveolens* and *Pterocarpus mildbraedii*
   (d) Several African genera have species in West Africa and the East African coastal forests e.g. *Mesogyne* (Moraceae) and *Isolona* (Annonaceae).

These would be explained if there had been former connections of rain forest between West and East Africa. Ideally, two connections could be postulated: a more ancient one to explain generic similarities, and a recent one to account for the species common to both areas. It is interesting to note that Moreau (1966) also postulates two former connections in this direction to explain speciation in birds.

3. Desert disjunctions. There are disjunctions at both the generic and specific levels between the northern and southern desert areas of Africa e.g. *Sporobolus spicatus, Echidnopsis* spp. There are similar disjunctions between the mammals (Kingdon, 1971) and the birds (Winterbottom, 1967).
4. Upland disjunctions. The similarities between the upland flora of Mt Cameroon and East Africa have already been mentioned; in addition, however, the separate peaks of East Africa are each an island of Afro-alpine vegetation in a sea of inhospitable areas. *Dendrosenecio* and *Lobelia* are good examples of this. In *Lobelia* there is well-marked vicarism: almost every mountain range has its own species. There are also disjunctions between taxa of the montane forests e.g. *Hagenia abyssinica, Podocarpus*. Evidence from distribution of birds (Moreau, 1966) and mammals (Coe, 1967) suggests former connections between mountain faunas and the botanical evidence supports this idea.

3.4 MODERN POLLEN RAIN

A modern pollen diagram from montane East Africa (*Figure 3.5*) has been compiled (Flenley, 1973) from the work of numerous authors. Samples have been placed in groups depending on the altitudinal zone or belt in which they are said to have been collected rather than in a strictly altitudinal sequence. Different authors used different pollen categories and different methods of calculation, so for uniformity all data were recalculated as percentages of total dry land pollen and spores. Additional samples from Kigezi (Hamilton, 1972) were omitted, as they show variation chiefly in proportions of swamp elements rather than of dry land types. Further samples from Mt Elgon were also omitted (Hamilton, 1972), as were some from Ethiopia (Bonnefille, in press).

It is clear that not all pollen types in the diagram behave in the same way as regards pollen dispersal. Some pollen types, e.g. *Afrocrania*, are rarely recovered outside the area of their production. This is probably due to their poor dispersal power, but since the results are all percentages and are therefore relative to each other, it could be due to excessive production by other pollen taxa. These types are therefore designated as displaying *low relative export* (Flenley, 1973). Some pollen types show *moderate relative export* while at the other extreme are pollen types which exhibit *high* or *very high relative export*, i.e. they are found in high percentages well outside the area of occurrence of the parent plants. The best example of this is *Podocarpus*, which occurs more abundantly (as a percentage) high in the Afro-alpine Belt and above, than in the Montane Forest Belt where the tree actually grows. Other pollen taxa which behave similarly in Africa are *Acalypha, Celtis*-type and Urticaceae (Hamilton, 1972). It is extremely interesting that this export of pollen is nearly always uphill; it is quite uncommon for pollen of species from above the forest limit, for example, to be detected below it.

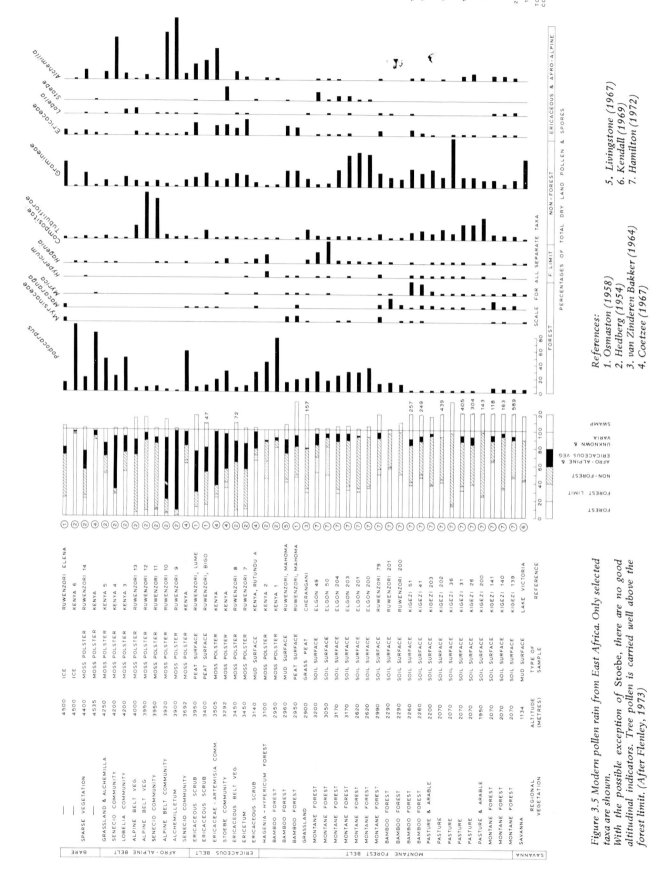

Figure 3.5 Modern pollen rain from East Africa. Only selected
taxa are shown.
With the possible exception of Stoebe, there are no good
altitudinal indicators. Tree pollen is carried well above the
forest limit. (After Flenley, 1973)

References:
1. Osmaston (1958)
2. Hedberg (1954)
3. van Zinderen Bakker (1964)
4. Coetzee (1967)
5. Livingstone (1967)
6. Kendall (1969)
7. Hamilton (1972)

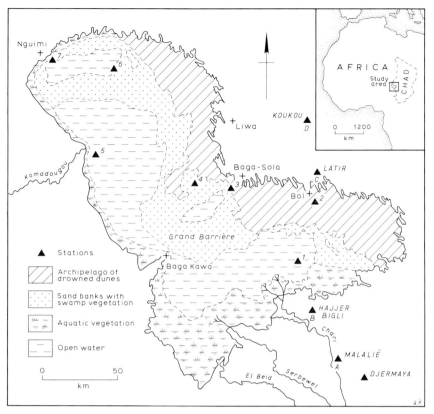

Figure 3.6 Lake Chad, showing stations for collection of modern pollen rain samples. (After Maley, 1972)

Greater production by the forest and better pollen dispersal because of the height of the trees are probably two reasons for this (Hamilton, 1972), but meteorological conditions could also be important. Winds on tropical mountains are usually up-valley during the day and down-valley at night. Presumably most anemophilous pollen is released during the day when lower humidity causes the anthers to open (Percival, 1950). A gentle breeze of a few kilometres per hour could carry pollen a long way uphill during the afternoon. Deposition would presumably occur in the afternoon or evening rainfall so common in the tropics, or in the still evening air. Thus little pollen would remain airborne to be carried down in the night-time wind. Upward pollen export is well known in temperate regions such as the Alps (Rudolph and Firbas, 1926).

Although the interpretation of any pollen value must depend on the remainder of the spectrum and on the environment of collection, some generalisations about particular pollen types can be made. Notes based chiefly on the conclusions of Hamilton (1972) and Livingstone (1967) are given in Appendix 1.

It will be apparent from this list that montane East Africa is poorly provided with pollen types which make really good indicators of ecological conditions.

Modern pollen rain studies in the lowlands are unfortunately very few; however, a study in the region of Lake Chad (Maley, 1972), although beyond the 10°N line of latitude, is of interest. The samples fall into two groups, a group chosen to demonstrate variations in pollen assemblages accumulating in different parts of the lake, and a group consisting of a

N–S transect across the whole area (*Figure 3.6*). The pollen types were classified according to the phytogeographical elements to which their producers belonged. This resulted in four elements being recognised: 1. sahelian, 2. sahelo-sudanian, 3. sudanian and sudano-guinean, 4. multi-regional. In addition the pollen taxa were grouped in ecological and taxonomic terms:

1. Herbaceous plants
2. Gramineae
3. Cyperaceae
4. Total trees
5. Regional trees
6. Combretaceae
7. Aquatic plants
8. *Aeschynomene elaphroxylon* (Papilionaceae), a swamp tree
9. Allochthonous plants (southern)
10. Allochthonous plants (northern).

The results are plotted as values (on a semi-logarithmic scale) relative to the most southerly station in each group (*Figure 3.7, 3.8*). It is interesting to note that the Allochthonous southern group is most abundant in the samples from the southern part of the lake, that is in the area where the Chari river enters the lake. This group is relatively rare in ponds (samples B and C), so is presumably brought to the lake mainly by river transport. In the regional transect (*Figure 3.8*) the increase in the Sahelian element and the decline of other elements towards the north accords well with the present vegetation of the region. A few long distance

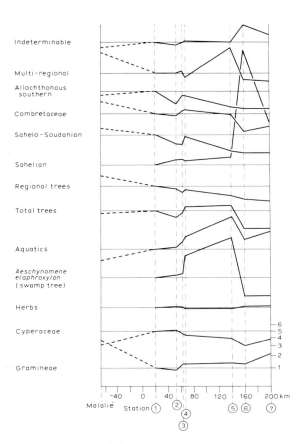

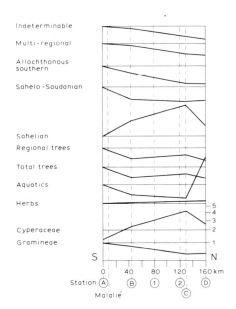

Figure 3.7 Modern pollen rain from Lake Chad. Transect of samples across the lake. Values are plotted relative to the value at Station 1, on a semi-logarithmic scale. (After Maley, 1972)

Figure 3.8 Modern pollen rain from Lake Chad. Regional N–S transect. Values are plotted relative to the value at Station A, on a semi-logarithmic scale. (After Maley, 1972)

pollen grains occur e.g. *Artemisia* and *Erica arborea*. These may have been derived from the Tibesti Mountains, 800 km to the north. But the overall results suggest that long distance dispersal is not sufficiently serious to obscure the true pattern. Studies of modern pollen rain in Ethiopia (Bonnefille, 1969, 1972, in press) will be considered in direct relationship to the fossil samples in *Section 3.5*.

The general conclusion from all these modern pollen rain studies is that, with care, it should be possible to interpret African pollen diagrams in terms of former vegetation. Particular care must be taken, however, to allow for the part played by far-travelled pollen, especially that moved uphill by mountain winds, and that brought into lakes by rivers.

3.5   THE FOSSIL POLLEN EVIDENCE

There are good reasons, which will shortly appear, for discussing separately the highland and lowland evidence. Since tropical pollen analysis actually began in highland Africa, with Hedberg's (1954) and Osmaston's (1958) reconnaissances, it is perhaps appropriate to consider the highland evidence first.

3.5.1   THE HIGHLAND EVIDENCE

The longest continuous sequence of deposits which has yet been studied is that from Sacred Lake at 2400 m on the slopes of Mt Kenya (Coetzee, 1967; van Zinderen Bakker and Coetzee, 1972). The mountain is an isolated extinct volcano of Pliocene age, rising to 5200 m, and bearing permanent snow (*Figure 3.9*). Sacred Lake lies in a small crater on the northeast side of the mountain, surrounded by montane forests of humid type. The forest limit is at least 500 m above it at about 2900–3000 m. The pollen diagram is from a core over 10 m long, mostly a fine detritus mud, collected below 4.5 m of water. A radiocarbon date at the base gave an age of 33 350 ± 1000 B.P. Some of the pollen types recovered from the core are shown in *Figure 3.11*. The diagram, of which a summarised version is shown in *Figure 3.10*, may be considered in nine zones.

The earliest zone, *R*, shows high values for shrubs of the Ericaceous Belt (including Compositae) and for Gramineae. *Hagenia* is quite well represented, as are Montane Forest Belt types and *Podocarpus*. Comparison with the modern pollen rain diagram will show that such spectra are typical of the Ericaceous Belt, and the upper part of the Montane Forest Belt. Possibly the forest limit was slightly below the lake at the time which means it was depressed by at least 500 m. *Cliffortia* and *Stoebe* make useful indicator taxa for this. This zone is, however, based on only a few samples from sandy material, and it would be unwise to place too much reliance on this evidence. In the succeeding zone, *S*, there is a gradual decline of pollen of Ericaceous Belt taxa, and to some extent of that of Gramineae. *Hagenia* pollen becomes more

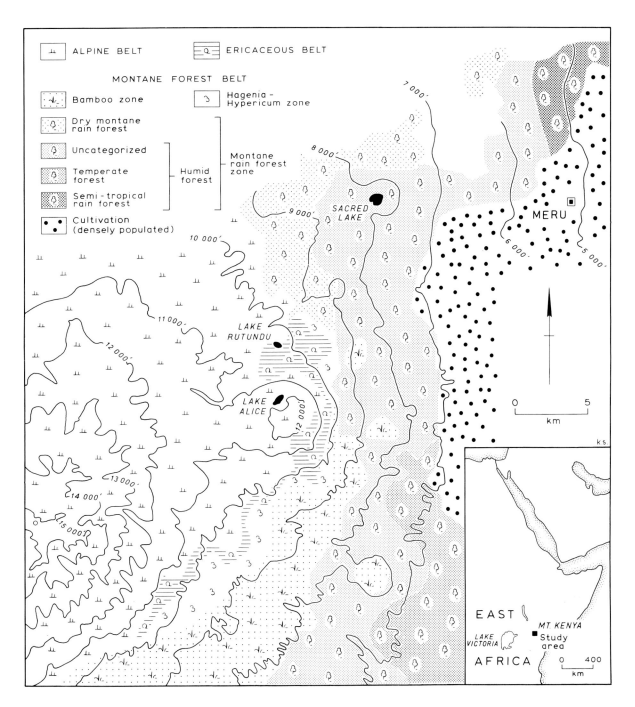

Figure 3.9 North-east side of Mount Kenya showing the vegetation belts and the positions of Sacred Lake and Lake Rutundu. (After Coetzee, 1967)

38

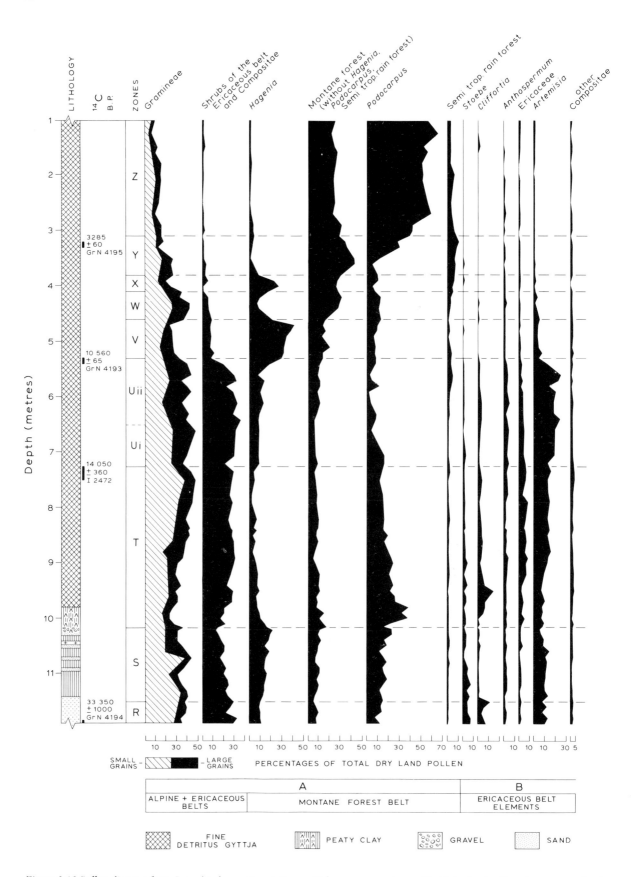

Figure 3.10 Pollen diagram from Sacred Lake on Mount Kenya. Values are percentages of total dry land pollen. A, taxa contributing to the pollen sum (which also includes unknowns); B, individual taxa of the Ericaceous Belt and Montane Forest Belt; only selected taxa shown. The successive dominance by elements of the Alpine and Ericaceous Belts, tree line and Montane Forest Belt is very clear. (After Coetzee, 1967)

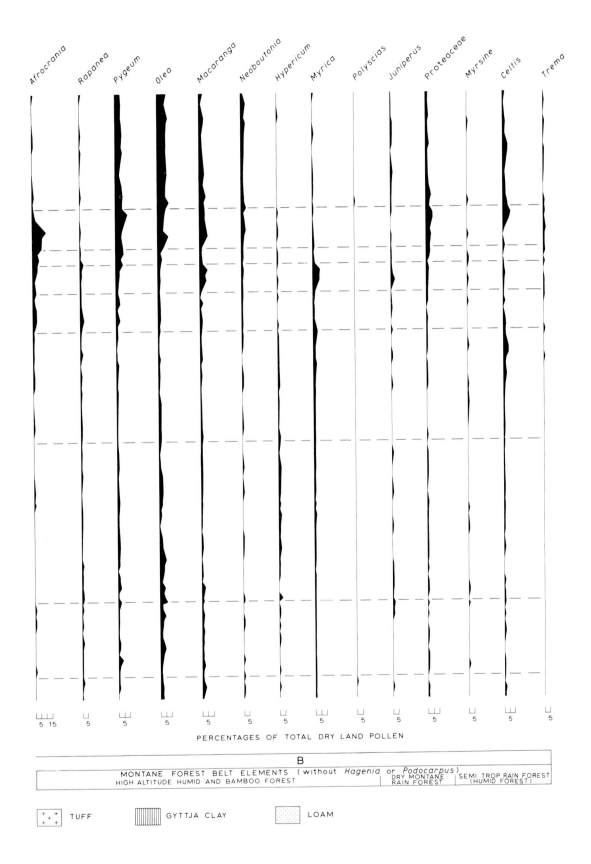

Afrocrania   Rapanea   Pygeum   Olea   Macaranga   Neoboutonia   Hypericum   Myrica   Polyscias   Juniperus   Proteaceae   Myrsine   Celtis   Trema

5 15   5   5   5   5   5   5   5   5   5   5   5   5   5

PERCENTAGES OF TOTAL DRY LAND POLLEN

B

MONTANE FOREST BELT ELEMENTS (without *Hagenia* or *Podocarpus*)
HIGH ALTITUDE HUMID AND BAMBOO FOREST | DRY MONTANE RAIN FOREST | SEMI TROP RAIN FOREST (HUMID FOREST)

+ + +
+ + +   TUFF   GYTTJA CLAY   LOAM

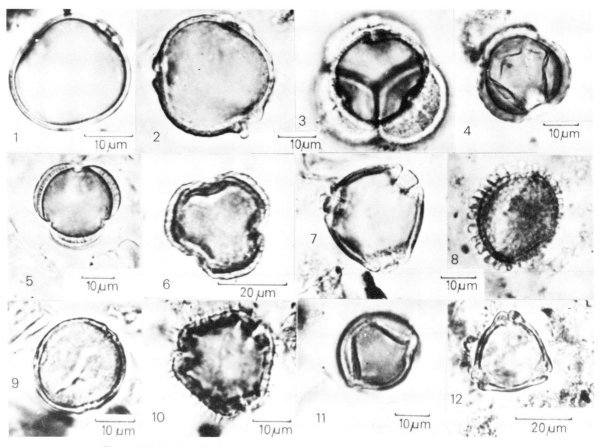

Figure 3.11 Fossil pollen from the Quaternary of East Africa. (After Coetzee, 1967)

1. *Gramineae*	5. *Artemisia afra*	9. *Celtis*
2. *Cliffortia*	6. *Anthospermum*	10. *Afrocrania*
3. *Ericaceae*	7. *Hagenia abyssinica*	11. *Pygeum*
4. *Stoebe kilimandscharica*	8. *Ilex*	12. *Myrica*

abundant, along with that of *Olea, Podocarpus* and other forest types. These trends have been interpreted as a gradual uphill migration of the forest limit until it coincided approximately with the altitude of the lake at the time represented by the end of the zone, an estimated radiocarbon age of c.26 000 B.P. The next zone, *T*, extends upwards to a radiocarbon date of 14 050 ± 360 B.P. Its chief feature is the marked decline of the *Hagenia* curve and corresponding rises in the curves for Gramineae and the Ericaceous Belt. Ericaceae and Gramineae percentages reach their highest values for the whole diagram. It is difficult to conclude otherwise than that the forest limit was again depressed below the lake at this time, and was probably a considerable distance below it. The presence of *Podocarpus* and other montane forest elements does not conflict with this, since they are capable of uphill transport (*Figure 3.5*). It is important to note the abundance of *Artemisia* pollen in this zone. This may well be an indicator of dry conditions (Coetzee, 1967; Hamilton, 1972).

In the next zone, *U, Artemisia* reaches its highest value in the diagram, and the zone is characterised by high values for the Ericaceous Belt pollen and moderate values for *Hagenia*. Many of the abundant Gramineae pollen of this zone fall into the larger size category, which means they could possibly be derived from *Arundinaria alpina*, the chief bamboo of the Bamboo Zone. Bamboo appears to contribute little to surface spectra, however (Morrison, 1961; Hamilton, 1972), so it remains quite likely that the fossil pollen originates otherwise. In this zone the forest limit was apparently once again close to the lake, and conditions may have been rather dry. Minor oscillations of the forest limit may possibly be indicated. The zone ends with a radiocarbon age of 10 560 ± 65 B.P.

The boundary between zones *U* and *V* is the most significant in the whole diagram. At this level there is a sharp increase in pollen of *Hagenia* and montane forest. The latter group includes pollen of *Afrocrania, Pygeum* and *Neoboutonia* which are types of low relative export (Hamilton, 1972; Flenley, 1973), and were therefore probably growing nearby. *Hagenia* reaches its maximum for the whole diagram, and the inescapable conclusion is that the *Hagenia* forest of the forest limit was immediately adjacent to the lake. The presence of *Afrocrania* indicates moist conditions.

In the following zone (*W*) the montane forest completely surrounded the lake, as demonstrated by the high values for montane forest pollen, and the decline in *Hagenia*.

Zone *X* represents a brief interval, possibly around c. 5800 B.P. and lasting perhaps 1000 years, when the forest limit appears to have moved down again.

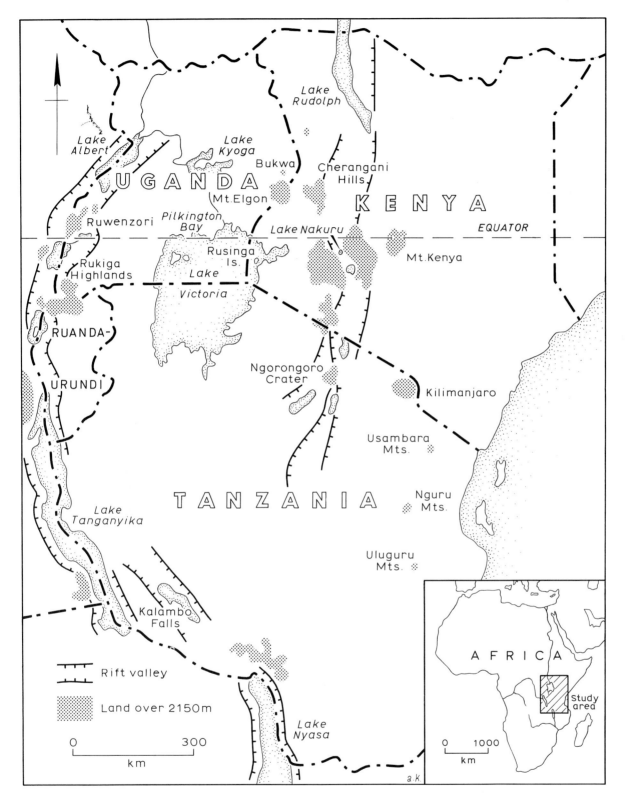

Figure 3.12 East Africa, showing locations of palynological studies and some other areas mentioned in the text. (After Lind and Morrison, 1974)

Certainly there is a pronounced resurgence of the *Hagenia* pollen curve in this zone.

Zone *Y* is again dominated by the pollen of montane forest species, and *Hagenia* pollen is almost absent. *Olea* and *Celtis* pollen, both types of high relative export, are abundant, and *Podocarpus*, which is of very high relative export, becomes abundant during the zone. The abundance of these, along with montane pollen types such as *Neoboutonia* suggests that Bamboo forest surrounded the site. The abundance of *Afrocrania* still indicates a moist climate.

The final zone, *Z*, shows such high values for *Podocarpus* and other montane forest elements that it seems likely *Podocarpus* was actually present in the nearby forest at this time. *Podocarpus* and *Olea* are characteristic of the dry type of montane forest which is very extensive today above and north of the lake, although the lake itself lies in humid forest (*Figure 3.9*).

The overall conclusion from this pollen diagram (in so far as one can draw conclusions from a single diagram) must be that there have been pronounced variations in the past vegetation of the mountain, along both the altitudinal axis of variation and the wet–dry axis of variation. If one assumes the usual relationship between altitudinal zonation and temperature, this would be evidence of both thermal and hydrologic change. Strongly generalising, the period c.26 000–14 000 B.P. appears to have been both colder and drier than the present, while the last 10 000 years appear to have resembled the present day. Other periods were apparently intermediate. It is important to realise, however, that these interpretations are based on a former vegetational pattern which was probably at least as complex as the present one. Some of the minor changes recorded in the diagram between different zones in the same belt are probably not related to climatic changes directly, if at all. It is also quite possible that the species were assorted into different communities formerly (Livingstone, 1975). In addition it is particularly true in Africa that human disturbance as a possible factor can never be ignored in vegetational history.

All these considerations make it essential that confirmatory evidence, in the form of additional pollen diagrams, should be discussed. Fortunately a number of such diagrams are now available from the sites shown in *Figure 3.12*. We should not expect, of course, that pollen diagrams from different sites would show identical results, but there should be a broad parallelism. This might be of two kinds: a 'regional parallelism' (von Post, 1916) between sites at similar altitudes in different areas and an 'altitudinal parallelism' (Flenley, 1973) between sites at different altitudes in the same area. Thus the pollen diagram from Lake Rutundu at 3140 m, also on Mt Kenya (*Figure 3.13*) suggests Alpine Belt vegetation in its earliest phase, ending at an inferred age of 14 000 B.P. This is the period when Sacred Lake was surrounded by Ericaceous Belt vegetation. After 14 000 B.P., when Sacred Lake became surrounded by forest, Lake Rutundu was apparently surrounded by Ericaceous Belt vegetation. This not only helps to confirm the

date of vegetation change on the mountain, but also does not conflict with the idea that vegetation belts were arranged in the past in an altitudinal sequence similar to the present one.

The finest sequence of altitudinally spaced sites is that from the Ruwenzori investigated by Livingstone (1967). Five episodes in the glacial history of the area are recognised by Osmaston (1965):

1. Lac Gris, represented by small fresh frontal moraines at about 4200 m, only a few hundred feet below the glaciers. Suggested age 100–700 years.
2. Omurubaho, represented by small scarcely eroded frontal moraines at 3600–3900 m. Suggested age 10–15 000 years.
3. Lake Mahoma, represented by large converging lateral moraines, only slightly eroded. These moraines extend down to 1740 m and a radiocarbon date of 14 750 ± 290 B.P. from the base of the Mahoma core provides a minimum age for deglaciation (Livingstone, 1962). The age of the moraines is therefore estimated at 15–20 000 years.
4. Rwimi Basin, represented by large U-shaped glacial valleys extending slightly beyond the limits of the Lake Mahoma moraines. Suggested age perhaps c.100 000 years.
5. Katabarua, represented by long frontal moraines on the east of the range and by extensive sheets to the north and west. The till is much weathered. Suggested ages greater than 100 000 years.

On the Ruwenzori the Alpine Belt extends from about 3000 m to the summits. Its four principal communities are *Dendrosenecio* forest, *Helichrysum* scrub, Alchemilletum and *Carex runssoroensis* bog. The Ericaceous Belt forms a wide irregular girdle from about 2700–3900 m or a little above. Osmaston delimits a *Rapanea–Hagenia* Zone (replacing the usual *Hagenia–Hypericum* Zone) between the bamboo and the heather (Langdale Brown *et al.*, 1964). *Hypericum* is apparently not abundant on the Ruwenzori. The Montane Rain Forest Zone extends from 2050–2400 m on the east side of the mountain, and from 1800–2300 m on the west. Much of the lower forest has been cleared by man, but lowland rain forest is present on the northwest side.

The lowest site studied is Lake Mahoma which, at 2900 m, is believed to be the lowest altitude glacial lake in equatorial Africa. The lake lies in the *Rapanea–Hagenia* Zone. A core 6 m long was raised from a water depth of 9.5 m. The radiocarbon age of the lowest organic part of the core, 14 750 ± 290 B.P. provides an important date for deglaciation of the area. The pollen diagram has been recalculated by Hamilton (1974) to exclude local pollen types and types of very high relative export, which makes it easier to interpret. The original diagram of Livingstone (1967) calculated as percentage of the total pollen and spores is presented here, however (*Figure 3.14*), for ease of comparison with those at higher altitude in the same sequence. Livingstone divided the diagram into two main zones and a number of sub-zones. Zone I (approx.

MT. KENYA : LAKE RUTUNDU

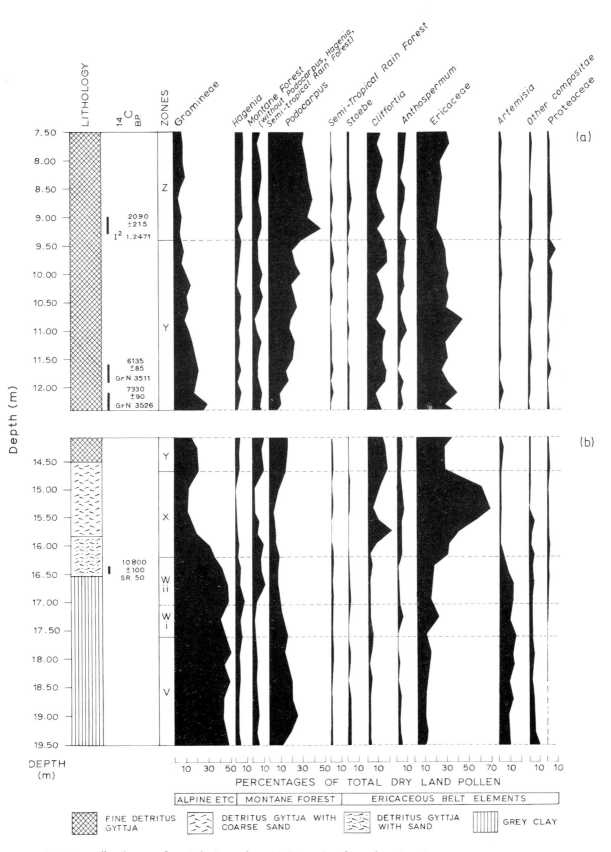

Figure 3.13 Two pollen diagrams from Lake Rutundu on Mt Kenya, E. Africa, plotted on the same scales. There is probably a slight overlap in stratigraphy between the two diagrams. Only selected taxa are shown. Values are percentages of total dry land pollen. There is clear evidence of movement of the forest limit. (After Coetzee, 1967)

44

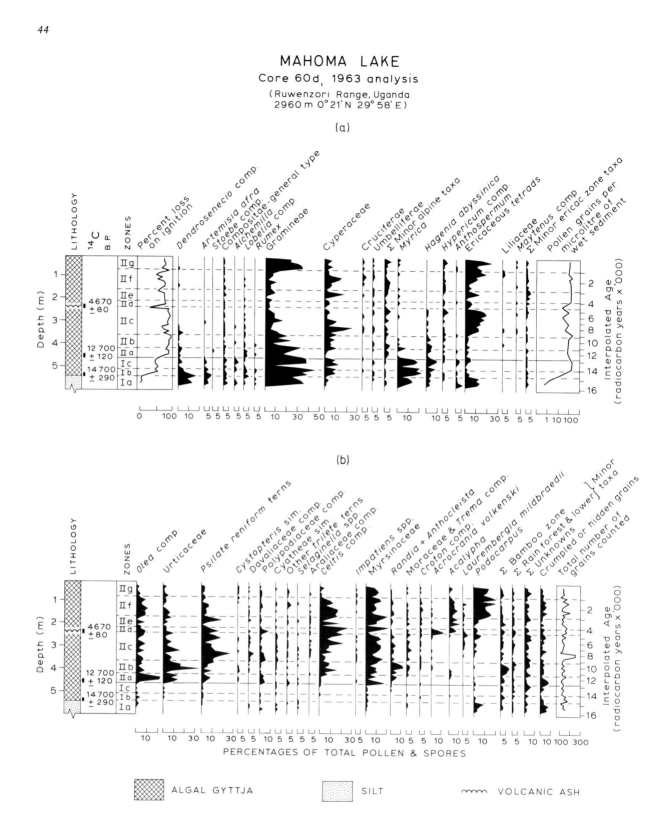

MAHOMA LAKE
Core 60d, 1963 analysis
(Ruwenzori Range, Uganda
2960 m 0°21'N 29°58'E)

(a)

(b)

PERCENTAGES OF TOTAL POLLEN & SPORES

ALGAL GYTTJA   SILT   VOLCANIC ASH

Figure 3.14 Pollen diagram from Lake Mahoma, Ruwenzori Range, Uganda (2960 m, 0° 21'N, 29° 58'E). Values are percentages of total pollen and spores.

(a) taxa important in the Alpine and Ericaceous Belts. 'Σ Minor alpine taxa' and 'Σ Minor ericaceous taxa' are composed of minor traces of pollen from plants that reach their upper altituduinal limits in the Alpine and Ericaceous Belts respectively.

(b) taxa important in the Montane Forest Belt and below it. 'Σ Bamboo Zone' and 'Σ Rainforest and lower' are composed of minor traces of pollen from plants that reach their upper altitudinal limits in the Bamboo Zone or below it, respectively. This diagram may be compared with Figure 3.18 which is from almost the same altitude, but in a much drier area (After Livingstone, 1967)

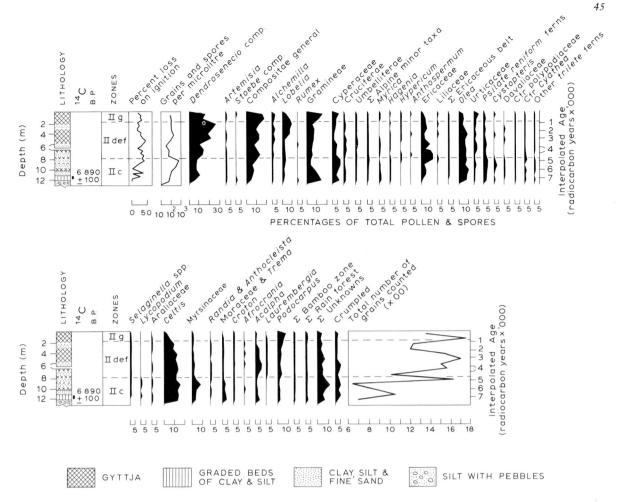

PERCENTAGES OF TOTAL POLLEN & SPORES

Figure 3.15 Pollen diagram from Kitandara Lake, Ruwenzori Range, Uganda (3990 m, 0° 21'N, 29° 53'E). Pollen counts were combined by metre intervals and the values shown are percentages of total pollen and spores. (After Livingstone, 1967)

Figure 3.16 Lake Kitandara, Ruwenzori, East Africa. The lake lies at 3990 m in the Ericaceous Belt. The vegetation is dominated by giant lobelias, with flowering spikes (Lobelia sp.) and giant ragworts (Dendrosenecio sp.). (Photograph by J. Harrop)

*Figure 3.17 Lake Bujuku, Ruwenzori, East Africa. The snow-covered peaks behind are twin peaks of Mt Stanley (5109 m). The lake, at 3920 m is in the Afro-alpine Belt and was the site for a pollen diagram by Livingstone (1967) covering the last 3000 years. The vegetation consists chiefly of the giant ragworts (*Dendrosenecio *spp.) and sedges. (Photograph by J. Harrop)*

15 000–12 900 B.P.) has high percentages for Gramineae, *Dendrosenecio*, Ericaceous tetrads, *Myrica* and (in Zone Ia) *Podocarpus*. The values are also higher in Zone I than in Zone II for *Artemisia, Stoebe, Alchemilla, Lobelia, Anthospermum* and *Hagenia* (the last reaching a peak in Zone Ic). These values are consistent with the modern pollen rain from the upper Ericaceous Belt or lower Afro-alpine Belt. That the Gramineae of Zone I are really of Afro-alpine affinities rather than of lower altitude savannah affinities has recently been confirmed by examination of leaf-cuticle fragments from the core (Palmer *et al.*, in press). The presence of *Podocarpus* need cause no surprise, since it is of high relative export. Likewise the presence of *Hagenia* (and Myrsinaceae) does not necessarily indicate the proximity of the *Hagenia–Rapanea* zone, at least until Zone Ic. *Myrica* is something of a problem. In the absence of swamps on which *M. kandtiana* might have grown, it seems likely *M. salicifolia* was responsible. This suggests that the forests below the lake were of a drier type than exists on the mountain today. Dryness around the lake is also suggested by the presence of *Artemisia*. Dry conditions in the lowlands are also suggested by the virtual absence of *Celtis*-type pollen in the record, since this would be expected to be carried up into the mountains if it had been available.

Zone II presents a great contrast with Zone I. Zones II a and b can be regarded as successional, and the rest of the zone is characterised by high values for Myrsinaceae, *Celtis, Olea,* Urticaceae and Pteridophyta. Gramineae values are relatively lower than before, except in the final sub-zone, IIg. These results suggest the presence of moist montane forest around the lake, probably bamboo forest. The proximity of forest is confirmed by the high value for *Afrocrania* in sub-zone IId. There is a striking rise in *Podocarpus* values in sub-zone IIf (c.3000 B.P.). This could represent the actual arrival of the species on the Ruwenzori, or perhaps its encouragement due to a minor climatic fluctuation. The parallel with the rise of *Podocarpus* about 4000 B.P. in the Sacred Lake diagram is interesting. Another important change in Zone II is the rise of *Acalypha*, starting about 4500 B.P. This is accompanied by a decline in *Celtis*-type, and much later by a rise in Gramineae. These findings are consistent with clearance of the lowland forest by man, although minor climatic dessication cannot be altogether ruled out as an alternative explanation.

We should expect that some of the above changes — those due to pollen types of high relative export — would show up in pollen diagrams from other altitudes on the Ruwenzori, while others — those essentially local in nature — might be reflected in parallel but

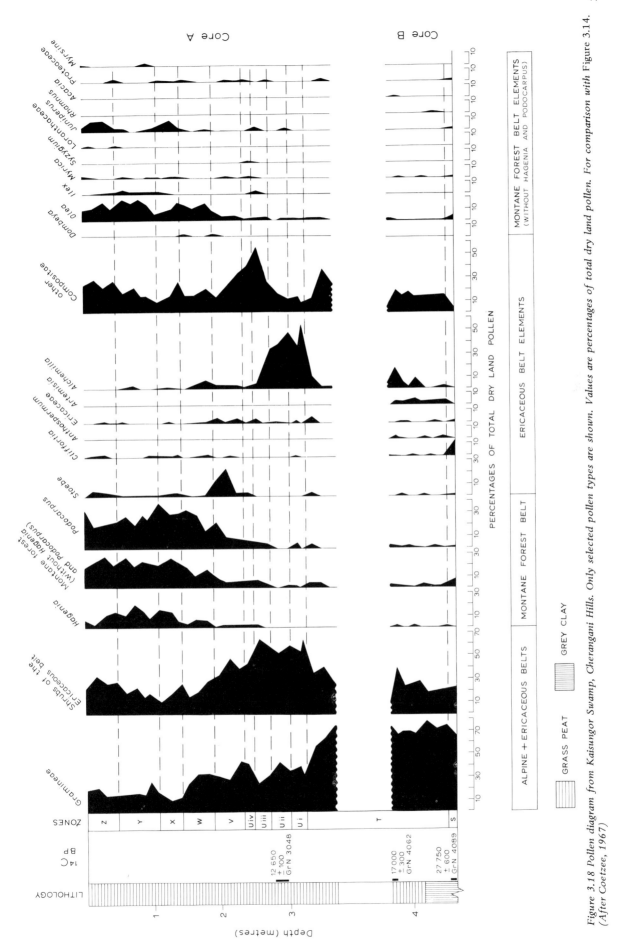

Figure 3.18 Pollen diagram from Kaisungor Swamp, Cherangani Hills. Only selected pollen types are shown. Values are percentages of total dry land pollen. For comparison with Figure 3.14. (After Coetzee, 1967)

different changes. The diagram from Kitandara Lake at 3990 m (*Figures 3.15* and *3.16*) allows us to test the idea. The site is in the area of the Omurubaho glacial phase and is correspondingly younger than Lake Mahoma; the oldest organic deposit gave an age of 6890 ± 100 B.P. Most of the Kitandara catchment is covered today by Afro-alpine vegetation, and the pollen diagram suggests this may have been so for the last 7000 years — for instance the high values for *Dendrosenecio, Lobelia* and Gramineae are typical of Afro-alpine spectra. The long distance pollen curves are very similar to those from Lake Mahoma. *Acalypha* and Gramineae show rises in the expected places, and *Celtis* declines. The rise of *Podocarpus* is less marked but still occurs. These regional changes are also seen in the short diagram from Bujuku Lake (*Figure 3.17*) at 3920 m (Livingstone, 1967).

An interesting regional comparison can be made between the diagram from Lake Mahoma at 2960 m and the diagram from Kaisungor Swamp at 2900 m in the Cherangani Hills (van Zinderen Bakker, 1962, 1964; Coetzee 1967). The Cherangani Hills have a very much drier climate than the Ruwenzori, and the present vegetation around the swamp is a grassland vegetation believed to have been formed by the clearance of dry montane forest. The pollen diagram from

here was in fact the first significant diagram to be published from East Africa, and a composite diagram from two cores (*Figure 3.18*) now extends back to 27 750 ± 600 B.P.

The pollen diagram may be considered in three main phases. In the first phase, c.27 750 to c.14 500 B.P., tree pollen is rare and Gramineae and Compositae (including *Dendrosenecio*) abundant. Vegetation of the Afro-alpine Belt may be postulated. When the diagrams are recalculated, excluding types of high relative export, high values of *Artemisia* and Chenopodiaceae are indicated, suggesting very dry conditions. The next two phases appear to reflect Ericaceous Belt and Montane Forest vegetation, in much the same way as the Mahoma diagram. There are important differences however. In the first place the type of forest indicated is a dry type rich in *Olea* and *Podocarpus* and with little *Myrsine*. Secondly the rise of forest pollen occurs at an inferred age of c.9000 B.P., which contrasts sharply with the date of 12 500 B.P. for the analogous change in the Ruwenzori at about the same altitude (Livingstone, 1967). It seems likely that the delay in forestation of the Cherangani Hills was related to their much drier climate, but the dates quoted are based on inference from too few radiocarbon assays to be sure as yet that the delay is real.

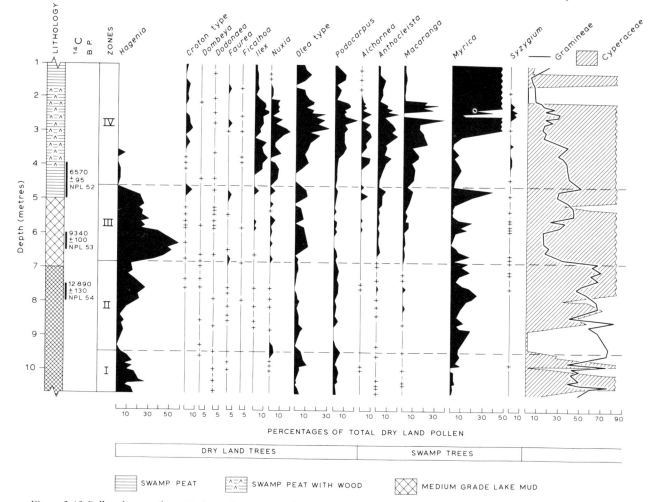

Figure 3.19 *Pollen diagram from Muchoya Swamp, Uganda. Values are percentages of total dry land pollen, i.e. swamp trees and aquatics are excluded from the pollen sums. For comparison with* Figure 3.10. *(After Morrison, 1968)*

Another useful comparison is between the diagram from Sacred Lake at 2440 m on Mt Kenya and that from Muchoya Swamp in the Rukiga Highlands, Kigezi, south-west Uganda (Morrison, 1961, 1968). The swamp lies at 2256 m altitude, in a present vegetation of bamboo forest. The oldest radiocarbon date obtained is 12 890 ± 130 B.P., but by extrapolation it has been suggested that the oldest sediments in the core were deposited about 24 000 B.P. Interpretation would be hampered by the abundant local swamp elements which are therefore excluded from the pollen sum. Hamilton (1974) interpreted the diagram (*Figure 3.19*) as follows:

Zone I (c.24 000 to c.17 000 B.P.). The boundary between the *Hagenia* Zone and the Ericaceous Belt occurs near the lake.

Zone II (c.17 000 to c.11 000 B.P.). The lake was surrounded by dry Ericaceous Belt vegetation.

Zone III (c.11 000 to c.6000 B.P.). Either *Hagenia* forest or bamboo forest surrounded the site.

Zone IV (c.6000 to c.1200 B.P.). The boundary between bamboo forest and montane forest lay near the site, which had now become a swamp.

The comparison of the Sacred Lake and Muchoya diagrams is extremely interesting. Before 12 000 B.P. the dating of the Muchoya diagram is non-existent, so any detailed attempt at correlation should be avoided. Around 11 000 B.P. the diagrams may be compared, and they show striking differences. For example the abundant *Artemisia* of Sacred Lake is completely missing in the Rukiga Highlands diagram where *Anthospermum* pollen is abundant. On the other hand the general trends of the *Hagenia* curves after 14 000 B.P. are rather similar at the two sites. *Olea* is abundant during the forest phase of both diagrams, but the chief accompanying tree-pollen types at Sacred Lake are *Pygeum, Macaranga, Neoboutonia* and *Afrocrania;* whereas at Muchoya they are *Nuxia* and *Ilex. Macaranga* is also abundant at Muchoya but probably largely originated from trees growing on the swamp there. The abundant *Myrica* and Ericaceae are also important swamp elements at the present day and presumably in the past. Most of these regional variations still await explanation in detailed ecological terms, but it is likely that moisture availability, temperature variations, human disturbance, edaphic factors, and historical accident all have contributions to make to the full ecological story.

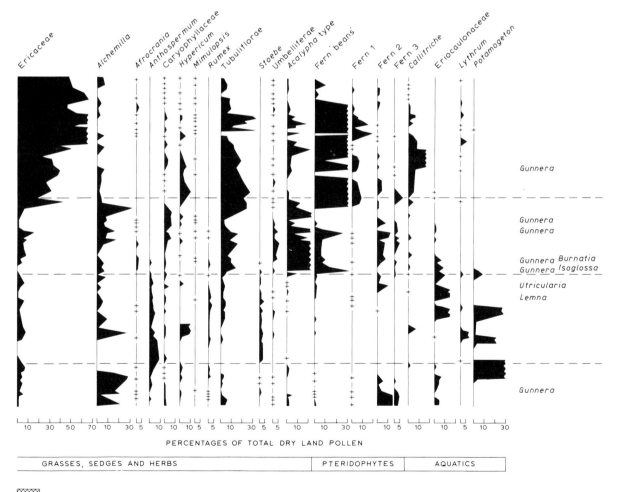

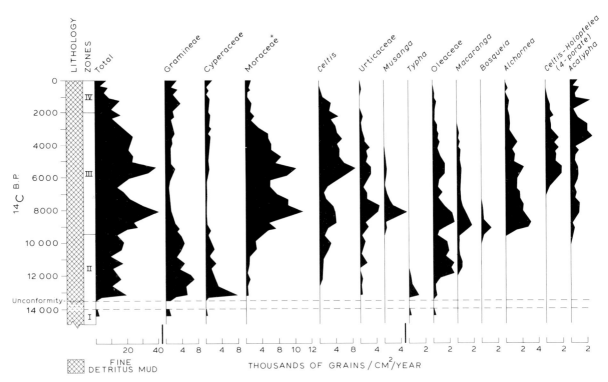

Figure 3.20 Pollen diagram from Pilkington Bay, Lake Victoria. Only major taxa are shown. Values are thousands of grains deposited per cm² per year. Heavy vertical lines draw attention to changes of scale. The broken lines indicate a discontinuity in deposition. The indication of vegetational changes is clear. (After Kendall, 1969)
* The Moraceae curve excludes certain genera, tabulated separately.

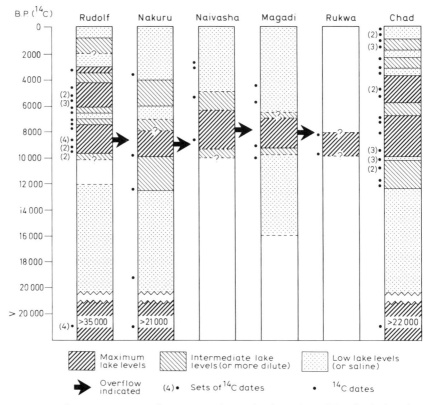

Figure 3.21 The Late Quaternary fluctuations of water level in African lakes that lack outlets at present. All the lakes apparently had a high water level in the period 10 000–8000 B.P. (After Butzer et al., 1972)

3.5.2 THE LOWLAND EVIDENCE

Several pollen diagrams from montane sites have led us to postulate that conditions were very dry before c. 12 000 B.P. on the East African plateau, and subsequently became more moist. Direct evidence for this was, however, lacking until Kendall (1969) produced his important analysis of a core from Pilkington Bay at the northern side of Lake Victoria at 1130 m. This core, 18 m long, was radiocarbon dated in no less than 28 places, and the base of the core is about 15 000 years old. This abundance of dates has permitted accurate sedimentation rates to be established, so that fossil pollen values can be expressed in absolute as well as relative terms. As it turns out, sedimentation rate is not all that variable, and absolute and relative diagrams do not differ too greatly. Only the absolute diagram, therefore, is presented here (*Figure 3.20*). Four provisional zones are established. In the first of these, c.15 000 to c.14 000 B.P., pollen preservation is poor and the sediment is oxidised. Other evidence (chemical analysis and algal fossils) confirms that the lake was drying up

at this time, and a disconformity at c.14 000 to c.13 500 B.P. indicates a time when Lake Victoria must have dried to below its outflow level. The little pollen found is largely Gramineae and *Podocarpus*. Since the latter is likely to be long distance pollen, the vegetation suggested is a grassy savannah. Forest appears to have been absent or of very limited extent. In Zone II, c.13 500 to c.9500 B.P., there is evidence for a gradual change from a low lake level, with swamp indicators such as Cyperaceae and *Typha*, suggesting dry conditions, to higher lake levels (abundant *Nymphaea*) with accumulation of pollen of forest trees – forest Oleaceae and *Trema* (*Olea* and *Trema* are pioneers today) and later taxa of semi-deciduous and then evergreen forest. After a minor oscillation back to drier conditions c. 10 300 to c.9500 B.P., there is abundant evidence for mesic conditions allowing good forest growth during Zone III (c.9500 to c.2000 B.P.). From c.5000 or 6000 B.P., however, there are increases in pollen of taxa characteristic of semi-deciduous forest (especially *Holoptelea* and *Celtis*), suggesting a change to more seasonal or slightly lower precipitation. This correlates with

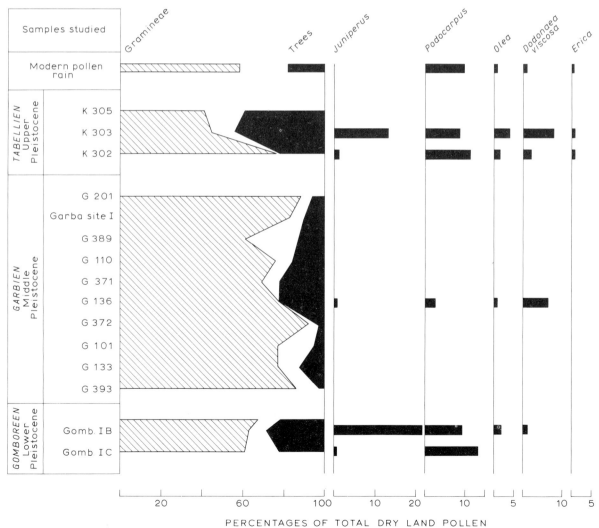

Figure 3.22 Pollen spectra from the Awash Valley, Ethiopia. Values are percentages of total pollen. (After Bonnefille, 1972)

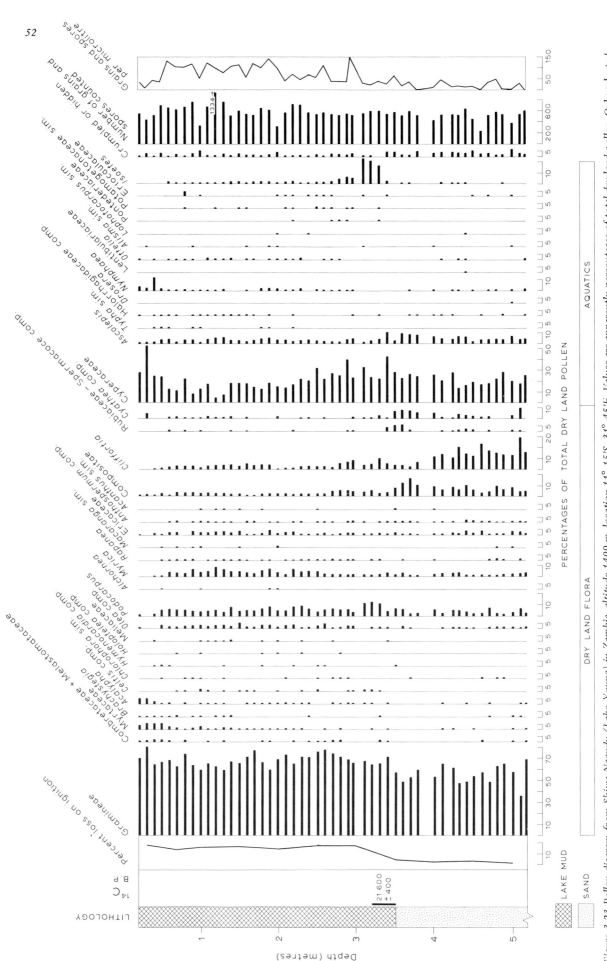

Figure 3.23 Pollen diagram from Shiwa Ngandu (Lake Young) in Zambia, altitude 1400 m, location 11° 15'S, 31° 45'E. Values are apparently percentages of total dry land pollen. Only selected taxa are shown. The diagram is remarkable for the lack of change in most of the last 22 000 years. (After Livingstone, 1971b)

evidence of declining water level in Lake Naivasha (Richardson and Richardson, 1972). The final 2000 years (Zone IV) is a time of increase of Gramineae and decline of forest pollen. This is probably the result of forest clearance, which will be considered in Chapter 7. In general, then, the Pilkington Bay core confirms earlier suggestions that the period just before c.12 500 B.P. was extremely arid in East Africa. The recent studies on levels of other East African lakes (Butzer *et al.*, 1972) spectacularly confirm this (*Figure 3.21*).

Information about similar fluctuations in vegetation earlier in the African Pleistocene are as yet very scanty, but a promising start has been made with the work of Bonnefille (1972) on the Upper Awash valley at 2000 m in Ethiopia. A number of analyses from separate sections allowed the construction of a composite pollen diagram (*Figure 3.22*) which shows distinct variations in the proportions of tree pollen and Gramineae pollen. Similar variations were also found in the results from the Omo valley at 400 m based on pollen and macro-remains (Bonnefille 1973, in press). These results show the way for future studies of long sequences from Africa.

It is not to be expected that a regional climatic change will necessarily be reflected in vegetational changes of equal magnitude everywhere; indeed the very reverse would be reasonable. In addition, many changes in tropical vegetation types may not be recorded in pollen diagrams due to the poor production of some tropical vegetation types. These two factors make it not surprising that there should exist a pollen record from tropical Africa which shows virtually no major changes in the last 22 000 years. This is the remarkable diagram from Shiwa Ngandu (Lake Young) at c.1400 m in Zambia (Livingstone, 1971b). The lake lies, at 11°S latitude, in *Brachystegia* (miombo) woodland, a vegetation type rich in Gramineae. Despite the large number of pollen types recognised, the diagram (*Figure 3.23*) shows few changes. There is a decrease in abundance of Compositae and *Cliffortia* around 3–4 m depth, but this coincides with a stratigraphic break, so it should be discounted. The only change with likely ecological meaning is the reduction in many tree pollen taxa around 0.5 m (perhaps c.3000 B.P.). This coincides approximately with the entry of Iron Age agriculture into sub-Saharan Africa (Clark, 1962). It is particularly unfortunate that *Brachystegia*, which dominates the present vegetation, never rises above 1% in the pollen record, and its pollen curve cannot therefore be interpreted adequately.

Turning now to the other side of the equator, we have a good deal of evidence about the water level of Lake Chad, but this is indirect as regards vegetation. Direct evidence is limited to pollen records from Lake Chad covering the period 700 A.D. to present (Maley, 1973) and to the sequences from the Saharan Mountains further north. Although the latter are outside our limit of study, it is of interest to note that van Campo *et al.* (1967) and van Campo (1967) concluded from a study in the Hoggar massif that contacts between the Boreal floras and humid tropical floras must have occurred intermittently throughout the

Pleistocene, although the abundance of desert elements in the pollen record at all times suggests that the Sahara has never ceased to exist. Evidence from the Ougarta mountains of the north-western Sahara suggests the occurrence of at least one humid interval during the Quaternary (Beucher, 1967). Similar evidence from the Tibesti (Cohen, 1970) was dated to early Holocene time.

3.6 CONCLUSIONS

When one considers that equatorial Africa is about as large as Europe, and that the number of pollen diagrams per unit area must be somewhat less than Europe had by 1930, it is clear that our present network of sites is inadequate for any firm conclusions. Nevertheless we have reached a stage where, at least for limited areas, there is a measure of consistency between the results, so that useful suggestions can be made.

3.6.1 VEGETATION

It seems clear from the evidence that there have been substantial changes in the vegetation during the Quaternary. In general, the vegetation in a given area during the period c.26 000 to c.14 000 radiocarbon years ago appears to have been of a type now found in areas with less total precipitation, or a more uneven seasonal distribution of rainfall, although there are significant exceptions to this (Livingstone, 1971b). In the period since c.8000 B.P., vegetation appears to have been very approximately as at present, except that human influence has become very significant in many areas. The period between c.14 000 and c.8000 B.P. can be regarded as broadly transitional, although in some areas it was exceedingly dry, at least until 12 000 B.P. In the mountains, but not the lowlands, there is also evidence of changes which can be interpreted as an altitudinal shift in major vegetation belts. These changes are summarised in *Figure 3.24*. It is true that Morrison (1966) has, very wisely, cautioned against too ready an acceptance of this idea, and it may well be that the hypothesis will be considerably modified in the future; for instance the present vegetation types may not be the same as those existing previously. But in general the evidence does not conflict with a widespread lowering of the limits of altitudinal belts by several hundred metres during the period c.26 000 to c.14 000 B.P., and their rise to present levels by c.8000 B.P. To attempt a more detailed reconstruction of the vegetation at this stage would be premature.

3.6.2 CLIMATE

Conclusions about palaeoclimatology are best left similarly as mere outlines. It may be suggested that the period c.26 000 to c.14 000 B.P. or 12 000 B.P.

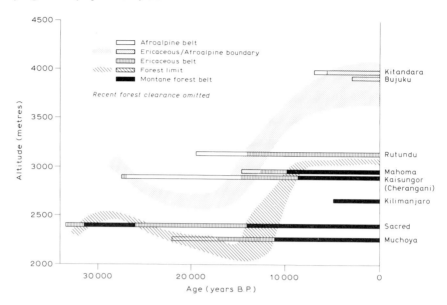

Figure 3.24 Summary diagram of Late Quaternary vegetational changes in the East African mountains. (Original)

was generally cooler and drier (compared with present climate), and the period since c.8000 B.P. very approximately as at present though with distinctly wetter phases in places. The period in between appears to have been transitional as regards temperature, but the evidence about hydrology is difficult to interpret. Moisture availability seems to have been very high in some places in the later parts of this period, but this does not necessarily mean precipitation was greater, since evaporation might be expected to have been reduced while temperatures were still below present values. It will be clear that this evidence is directly contrary to the old 'pluvial' theory in which wet periods were equated in time with the glacials of temperate areas. The evidence does not, however, conflict with the idea that a reduced global atmospheric circulation in glacial times led to reduced evaporation from the oceans, so that 'glacials' are roughly equivalent to 'interpluvials', at least near the equator.

3.6.3   BIOGEOGRAPHICAL PROBLEMS

It is perhaps worthwhile at this stage to check whether any of the biogeographical problems listed in Section 3.3 may hope for eventual solution from Quaternary reconstructions. The prospect for this now seems encouraging. The depauperate nature of the African forests *vis-à-vis* those of other tropical areas may now be seen to be possibly related to the extinctions caused in arid phases, which were perhaps more severe in Africa than elsewhere in the tropics (Richards, 1973). On the other hand, Laurent (1973) argued that the restriction of rain forest to refugia had led to increased diversity of the herpetofauna, rather than to depauperation. The disjunctions between West African forests and isolated forest patches in East Africa could be the result of plant migrations in times

of maximum wetness, possibly for instance c.9000 to c.7000 B.P. around Lake Victoria according to Kendall's evidence, followed by extinction of intermediate stations during arid phases. It seems likely that the Quaternary might have had several such wet phases at different times, thus accounting for floral similarities at various taxonomic levels.

The disjunctions between northern and southern deserts may similarly be explained as having originated during the equatorial dry phases for one of which we already have good evidence. It would be most satisfactory if the actual migrations of disjunct taxa could eventually be traced palynologically through the equatorial zone.

The disjunction and vicarism of montane taxa is much what would be expected to result from a former lowering of the limits of altitudinal belts. This would have reduced the barriers between mountain ranges, making dispersal more possible. Vicarism would result from isolation when altitudinal limits rose.

Most of these explanations are inadequate in two ways. Firstly, the migrations postulated have in most cases not been demonstrated palynologically or in any other way. Secondly, where there is evolution involved, for example in the formation of vicarious species, the time since the vegetational changes actually so far demonstrated may be inadequate. This does not mean the explanations are wrong in principle. It is likely that what has been demonstrated for the last 30 000 years is but the tail end of a series of cyclical events which have gone on for the two million or so years of the Quaternary. What is really needed now in Africa is the study of a few really long sequences. Fortunately such sequences, perhaps unbroken since the Miocene, in all probability do exist in the Rift Valley lakes. The field problems of obtaining these sequences, and the laboratory problems of studying them so as to obtain incontrovertible evidence are Africa's paleobotanical challenge.

4

The Quaternary Vegetation of Equatorial Latin America

4.1 INTRODUCTION

At the start of the Quaternary in the Americas, the Isthmus of Panama had been constructed and the Andes were already partially formed (*see* Chapter 2). The availability of montane habitats was already beginning to permit migration of temperate taxa southwards through Panama, and in all probability a similar migration from the southern to the tropical parts of South America was occurring. Lowland vegetation, in so far as it is known from the fossil record, was a rain forest of approximately present-day type.

4.2 PRESENT VEGETATION

A map of vegetation distribution (*Figure 1.5*), illustrates the tremendous areal extent of lowland rain forest at present. It occurs not only throughout the Amazon basin, but also on the Pacific side of the Andes, up to 1000 m altitude. This is the *Selva neotropical inferior* of Cuatrecasas (1958). Where climate is less generous with moisture the usual variety of semi-evergreen forests and xerophytic formations occurs, culminating in the cactus and thorn scrubs of dry areas. Soil factors also play their part here. For instance the free-draining nature of the calcareous soil on the island of Curacao probably encourages the prevalence there of a dwarf woodland of thorny leguminous shrubs with emergent columnar cacti, although the dry climate is clearly the major controlling influence.

The term savannah originated in Central America and at first meant any kind of low growing vegetation (Richards, 1964), and presumably the area of land supporting this. The areas usually now recognised as savannahs are essentially grass- and sedge-dominated though often with some trees and shrubs present. Several types of this have been described in the island of Trinidad (Beard, 1946), and there are very extensive patches in Guyana, especially the Rupununi-Rio Branco areas. Among the largest equatorial savannah areas are the Venezuelan and Colombian Llanos, first described by Humboldt (1852). The boundary between savannah and forest is often sharp, and 'gallery' forest (on the moist strips beside rivers) is common.

The altitudinal zonation of Andean vegetation is perhaps best known in northern Peru where it has been studied by Weberbauer (1914, 1922), in Colombia where Cuatrecasas (1958) and van der Hammen (1974) have described it, and in Venezuela (Hueck 1966). In Colombia the Andes split into three main parallel ridges, and altitudinal boundaries are somewhat lower on the Cordillera Occidental, nearest to the Pacific (*Figure 4.1*). The *Selva inferior* gives way at about 1000 m to the *Selva subandina* which contains less species than the lowland forest but is still rich in taxa of lowland affinities. Three genera that produce abundant pollen (*Acalypha, Alchornea* and *Cecropia*) do not occur above this zone. At around 2400 m the *Selva andina* begins (*Figure 4.2*). In most areas this is dominated by one or more species of *Weinmannia*, usually accompanied by *Ilex, Escallonia, Miconia, Hesperomeles, Podocarpus,* and in some

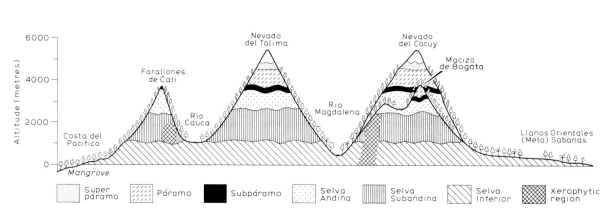

Figure 4.1 Simplified section across the three Andean cordilleras in Colombia, showing approximate zonation of vegetation. (After Cuatrecasas, 1958)

Figure 4.2 Cloud forests (of Weinmannia, Clusia etc.) in the Eastern Cordillera, Colombia, S. America, c.2800 m. The trees are completely covered by mosses and other epiphytes. (Photograph by T. van der Hammen)

Figure 4.3 Paramo-type vegetation, with Espeletia, above 3300 m in a boggy area between Tunja and Arcabuco, Colombia, South America. (Photograph by T. van der Hammen)

areas *Polylepis* and *Alnus. Polylepis* occurs especially at higher altitudes. In addition we must mention the oak genus, *Quercus*, which may dominate the forests in some areas (van der Hammen and Gonzalez, 1960).

The upper limit of forest is about 3200 – 3500 m (van der Hammen, 1974), or occasionally up to 3800 m (Cuatrecasas, 1958). Occasional forest patches are found up to 4000 m, dominated by *Polylepis* (Gonzalez *et al.*, 1966; Cuatrecasas, 1958). *Polylepis* occurs up to 4500 m in Venezuela (M.L. Salgado-Labouriau, personal communication).

Immediately above the forest is a transition zone called the Sub-paramo. This is analogous to the Ericaceous Belt of the African peaks and is similarly dominated by shrubs, although small trees are present. Important genera are *Vaccinium, Hesperomeles, Clethra, Miconia, Hypericum, Senecio, Weinmannia, Gaultheria* and *Tibouchina*.

The Sub-paramo is of varying width, with an altitudinal range usually around 200 m in any one area. Above it is the broad Paramo, reaching up to about 4100–4500 m. The Paramos are grassy meadows, where the most characteristic plants are species of the genus *Espeletia* (Compositae) (*Figure 4.3*). These have the pachycaul growth form, analogous with *Lobelia* and *Dendrosenecio* in Africa. Likewise, they also occur as vicarious species on different mountain ranges. Amongst the many herbs are species of *Gentiana, Halenia, Valeriana, Bartschia, Geranium, Plantago, Ranunculus* and *Paepalanthus* (van der Hammen, 1974).

Van der Hammen and Gonzalez (1960) report the occurrence in the Paramo of an Acaenetum dominated by the rosaceous *Acaena cylindrostachya*, usually on rocky soil with little humus, at altitudes between 3450 and 4300 m.

Above the Paramo there is a zone of almost bare ground, the Super-paramo. This occupies the 300–400 m (altitudinally) immediately below the permanent snow, which usually begins around 4300–4800 m. Frost action is severe, but a few plants survive, e.g. *Draba* spp. and *Senecio niveo-aureus*.

Rain-shadow effects are common in the mountains, and corresponding xerophytic vegetation (usually grass-dominated, sometimes with cacti) is locally prevalent.

Related montane zonations, with zones generally at lower altitudes because of the Massenerhebung effect, have been described from Trinidad (Beard, 1946) and Jamaica (Asprey and Robbins, 1953) but since there are no fossil data from these areas, these will not be considered here.

4.3   BIOGEOGRAPHICAL PROBLEMS

These problems may be considered under four main headings: mountain vegetation, lowland rain forest, savannah and mangrove.

The mountain problems are essentially similar to those of the African mountains i.e. disjunction, endemism, and vicarism. The Andes make continuous ranges, unlike the separate African peaks, so it is not surprising that these problems are less marked in South America, but they still exist, particularly in Super-paramo vegetation. Thus there is a relatively high degree of endemism in the Super-paramo of the Sierra Nevada del Cocuy (van der Hammen, 1974). Disjunction occurs between all the separate Paramo areas. *Espeletia* provides a good example. At a generic level there is disjunction between the various Paramo areas where *Espeletia* occurs. At a specific level, there is vicarism, for *Espeletia* occurs in many species, a proportion of which are restricted to individual Paramo areas, while others are of widespread occurrence.

If there is a problem associated with the lowland rain forest, it is its diversity. It may seem paradoxical to say this, when we have already regarded the African rain forest as abnormal for being less diverse than the South American, but the point is that there is as yet no widely agreed explanation for any particular degree of diversity in rain forest, so that diversity is always a biogeographical problem there. Certainly the Amazonian forest is exceptionally diverse.

It is perhaps worth noting at this stage that Haffer (1969) explained the diversity of the Amazonian bird fauna by suggesting that during the Pleistocene the forest must have been broken up into at least seven blocks in which speciation occurred. The same argument was used to explain diversity in other animal groups (Vanzolini and Williams, 1970; Simpson 1971; Brown, 1972) and in the flora (Prance, 1973).

The problems of the savannah in South America are well known (Beard, 1953; Cole, 1960; Anon, 1963; Hills, 1965; Labouriau, 1966; Ferri, 1971). What was its origin? Is it man-made or 'natural'? Is it maintained by climate, edaphic factors or fire? How do the different savannah areas come to have much the same species complement? It is relevant here to note that Eden (1974) concluded that savannah islands in southern Venezuela must be relics of more widespread savannah which existed formerly. The islands have survived through burning, when other areas became forested, but their floristic affinities suggest they were formerly in direct contact with each other. Similar savannah islands occur in the Amazonian forest (Spruce in Massart *et al.*, 1929).

4.4   MODERN POLLEN RAIN

There have been unfortunately far too few surveys of contemporary pollen deposition in Central and South America, although further large studies are under way (T. van der Hammen, personal communication).

The study by Muller (1959) of the paynology of contemporary Orinoco delta and shelf sediments is particularly interesting because it is one of the few studies anywhere in the world where transport of pollen by rivers and offshore currents has been considered. Muller indeed found that the Orinoco was transporting vast quantities of pollen. Perhaps the most remarkable example is that of *Alnus*, of which occasional grains were found although the source area is 800 km away and at an altitude of over 2000 m. A considerable amount of sorting must also have been in progress, related to the rate of settling of individual pollen grains. Large, heavy grains such as those of *Symphonia* (*Figure 4.4*) were scarcely found away from source areas in the delta and on the island of Trinidad. Small, light grains, however, such as *Rhizophora* (*Figure 4.5*), were readily transported. As a percentage, *Rhizophora* grains actually reach their maximum either next to the mangrove where they are produced or well out to sea where other grains are less abundant. Muller's survey also includes a number of surface samples in transects across different parts of the deltaic swamps (*Figure 4.6*). These illustrate principles similar to those demonstrated above. Thus *Rhizophora* pollen is almost ubiquitous, whereas the pollen of the other common mangrove genus *Avicennia* is almost restricted to occurrence in the mangrove vegetation. The palm genus *Mauritia* has pollen which is likewise almost restricted to its main source area, the 'Morichal'. *Symphonia* pollen is widespread in these samples, but that is because the parent tree is present in these very swamps. Van der Hammen (1963) examined a few surface spectra from British Guiana, and found that Muller's (1959) results were applicable there also.

A study of pollen dispersal in the Colombian Andes (van der Hammen, 1961b), was the first attempt at a large-scale trapping programme on an absolute basis for the tropics. Previous studies all gave percentage results, but the Colombian data can be expressed as grains per cm^2 per annum. This was achieved by establishing six stations on and above the Bogota plateau at elevations ranging from 2560–3450 m, and operating these for a year. At each station sticky slides were exposed, horizontally and vertically, for one- or two-weekly periods. The spectra obtained are interesting for several reasons. When they are shown plotted against the calendar (for example Station I, *Figure 4.7*) the results demonstrate the lack of seasonality characteristic of a tropical montane situation, although some taxa, for example *Miconia* and Ericaceae, still show clear seasonality. The annual catch at each station varied between 50 and over 500 grains per cm^2 per annum; these are very low values (Flenley, 1973), but this may be an artefact of the catching method used. If the results are expressed as yearly spectra they are of more direct application to interpretation of fossil results. A number of surface samples from other Colombian areas may then be included (*Figure 4.8*). For convenience the pollen types have also been grouped into trees and shrubs, Gramineae and *Acaena/Polylepis*. The pollen of *Acaena* and *Polylepis* is not distinguishable with certainty under the light microscope, although this can be done with scanning electron microscopy. The samples from the 'forest zone' show varying amounts of tree and shrub pollen depending on the proximity of cleared land. It seems clear that under fully forested conditions pollen of trees and shrubs would predominate here. In the 'alpine' Paramo zone, Gramineae pollen values tend to be very high, as are values for Compositae

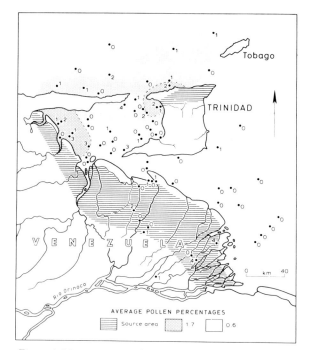

Figure 4.4 *Recovery of pollen of* Symphonia globulifera *from surface samples near the mouth of the Orinoco River. Values are percentages of total pollen excluding reworked types. This type is of low relative export. (After Muller, 1959)*

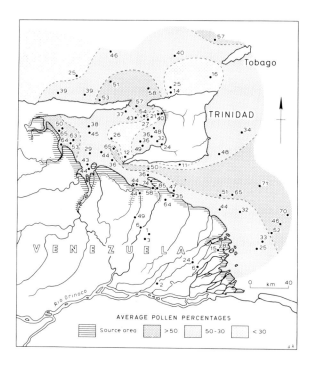

Figure 4.5 *Recovery of pollen of* Rhizophora *from surface samples near the Orinoco River. Values are percentages of total pollen excluding reworked types. This type is of very high relative export. (After Muller, 1959)*

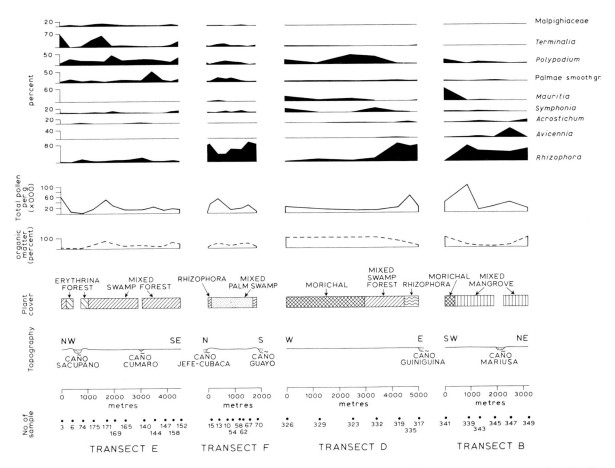

Figure 4.6 Pollen distribution in surface samples from transects across the Orinoco delta. Values are percentages of total pollen excluding reworked types. Some types are restricted to their area of production, but others are widespread. (After Muller, 1959)

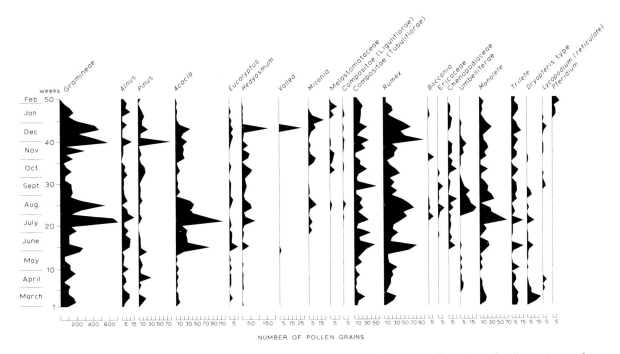

Figure 4.7 Modern pollen rain throughout a year at Bogotá, Colombia. Values are actual numbers of pollen grains caught on sticky slides. (After van der Hammen, 1961b)

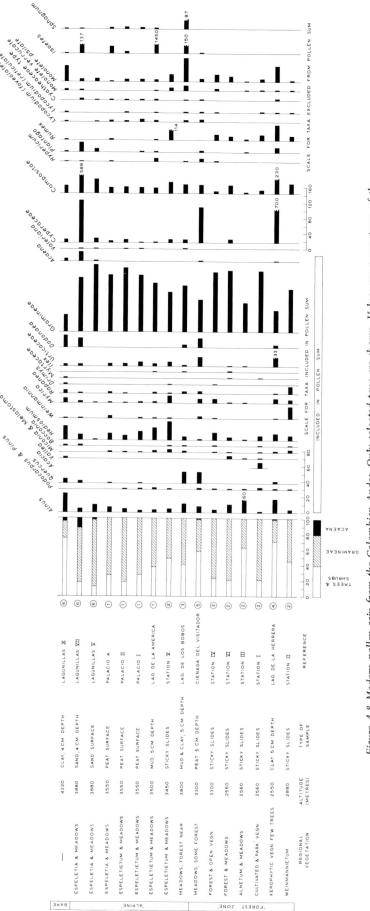

Figure 4.8 Modern pollen rain from the Colombian Andes. Only selected taxa are shown. Values are percentages of the totals for those taxa included in the pollen sum.

References:
1. van der Hammen and Gonzalez (1960);
2. van der Hammen (1961b);
3. van der Hammen (1962);
4. van der Hammen and Gonzalez (1965a);
5. van der Hammen and Gonzalez (1965b);
6. Gonzalez, van der Hammen and Flint (1966).

It is clear that forest pollen can be carried to high altitudes. (After Flenley, 1973)

pollen, probably *Espeletia*. In the bare Super-paramo, there is again a majority of pollen of trees and shrubs. This must reflect pollen carried uphill as in Africa. Here the main types which spread readily in this way appear to be *Alnus, Dodonaea, Hedyosmum, Quercus, Myrica* and *Rapanea*. These high altitude spectra are distinguishable from those in the 'forest zone' by several criteria. Firstly, the total pollen accumulation rate is usually very low; unfortunately this cannot be measured for fossil samples unless the sedimentation rate is known. Secondly, the values for Compositae are often high, whereas they may be low in forested areas. Thirdly, samples from the upper Paramo (above 3800 m) and the Super-paramo often have considerable values for the *Acaena/Polylepis* type. It is uncertain whether this originates from the Acaenetum of the Paramo, from *Polylepis* in the Andean forest, from isolated *Polylepis* forest patches in the Paramo, or from shrubby *Polylepis* spp. which exist in the Paramo (Schimper, 1903), but whatever its origin, it appears to be quite a good indicator of upper Paramo and Super-paramo situations.

Among the 'forest zone' samples, the single spectrum from the xerophytic vegetation at Laguna de la Herrera is distinct in many ways, particularly in its very high values for Cyperaceae and Compositae. General notes about the commoner pollen types are given in Appendix 2.

In the savannahs of Central Brazil, a study of modern pollen rain has been carried out by Salgado-Labouriau (1973). Pollen deposition was highly seasonal, by contrast with that in the Andean forests, and the spectrum was strongly dominated by Gramineae.

4.5   THE FOSSIL POLLEN EVIDENCE

The evidence of Quaternary vegetation in equatorial Latin America has been almost entirely derived from the studies of T. van der Hammen and co-workers. In a long series of papers they have revealed important facts about the former vegetation of the Colombian Andes, the coastal lowlands and the Amazon basin. The present survey depends heavily on a recent review by van der Hammen (1974).

4.5.1   THE COLOMBIAN ANDES

For most of the diagrams presented here, the pollen types have been placed in the following groups:

1. Pollen from taxa common in the Paramo (dominating elements Gramineae).
2. Pollen from taxa common in the highest zone of the Andean forest and shrub, or the Sub-paramo respectively.
3. Pollen from elements common in both the Andean and Sub-Andean forest.
4. Pollen from elements of the Sub-Andean forest.
5. Pollen from elements common in the tropical forest.

The allocation of taxa to these groups is given by van der Hammen *et al.* (1973).

In Chapter 2 we left the Andes half-formed, so to speak, and with a number of northern genera migrating down the Isthmus of Panama. The evidence for the next events comes from a truly remarkable series of deposits at the Sabana de Bogota, an old lake bed at 2600 m upon part of which the city of Bogota is built (*Figure 4.9*). The former lake was in an intermontane basin of tectonic origin, and has been infilling since Pliocene time. The oldest deposits of Pliocene age from the Sabana de Bogota (*Figure 4.10*) are dominated by pollen of lowland tropical elements. Even if we were to allow for a considerably warmer climate in the Pliocene, this would still indicate that the Bogota Lake stood at a much lower altitude formerly. The middle part of the Tilata formation, which is of Later Pliocene age, shows a dominance of Andean plus Sub-Andean elements. The species present suggest that, given a climate like today's, the lake was then at about 1500–2300 m. The Pliocene record includes the first occurrences of *Hedyosmum* and *Myrica* in South America. Presumably these (at least the *Myrica*) had recently migrated in from North America. After their first appearances they are present continuously in the record. The Late Pliocene phase already includes some Paramo pollen (probably from nearby peaks) and in the Early Pleistocene part of the sequence this expands to 50% or more although it is still poor in species. This implies a vegetation which is today found well above the Sabana de Bogota, and perhaps indicates an early cold phase of the Quaternary. The taxa of this primitive Paramo appear to have been Gramineae, *Polylepis/Acaena, Aragoa, Hypericum,*

Figure 4.9 Drilling equipment of the Colombian Geological Survey being used for the collection of undisturbed cores from the Pleistocene lake sediments of the Sabana de Bogotá, Colombia, South America. (Photograph by T. van der Hammen)

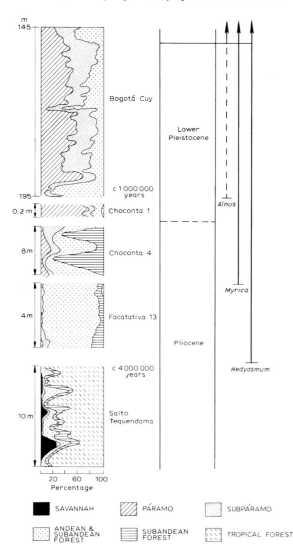

Figure 4.10 Summarised pollen diagram from the Pliocene and Lower Pleistocene of the high plain of Bogotá (Colombia), demonstrating the uplift of the area and the Early Pleistocene glacials and interglacials. (After van der Hammen, 1974)

Miconia, Umbelliferae, *Borreria*, *Jussiaea*, *Polygonum*, *Valeriana*, *Plantago*, Ranunculaceae, *Myriophyllum* and *Jamesonia*. Some of these presumably evolved from the surrounding tropical lowland vegetation, while others, for exampke *Plantago* and *Polygonum*, since they are of temperate affinities, probably arrived along the Andean ranges from north or south temperate regions. The forest elements were also being added to at this time, and *Alnus* makes its first sporadic appearance, probably also by migration across the mountains of the Isthmus during a phase of cool climate. A generalised reconstruction of the formation of the Andes and their vegetation belts is attempted in *Figure 4.11*.

The main Pleistocene sequence from the Sabana de Bogota is a continuous deposit of 200 m of sediments leading up to the present day. It is indeed a fortunate accident that the lake happened to reach a silted-up state during the Holocene, providing at least in some areas a complete sequence to the present day. The provisional diagram of the whole sequence (*Figure*

4.12) (van der Hammen and Gonzalez, 1964) shows two remarkable features. The first is the gradual augmentation of the montane flora, presumably as more taxa arrived by immigration. *Alnus* appears in this sequence only at the start of the Middle Pleistocene. The first appearance of *Quercus* is much later than that of *Alnus*. The possibility that this represents selective destruction in the less favourable sediments is unlikely, since after their appearance the pollen types appear regularly, whatever the nature of the sediment. Progressive enrichment was not confined to the forest flora, but also occurred in that of the Paramo, where, for instance, *Lycopodium* (two types), *Gunnera* and *Gentiana* are progressively added. The second feature is the impressive oscillation between zones dominated by trees and shrubs and those dominated by Gramineae and *Polylepis/Acaena*, in other words between forest elements and Paramo elements. There are at least two possible explanations for this. Either the changes represent real altitudinal shifts of vegetational zones – presumably in response to changes of climate – or they represent tectonic oscillations of the whole Sabana de Bogota basin. The continuation of tectonic movement in the area during the Quaternary is confirmed by the nature of the sediments in the Sabana de Bogota. Peat layers are present at intervals throughout the 200 m section. If the basin were 200 m deep at the start of deposition, an exceedingly arid climate would be needed to dry up the basin sufficiently for peat formation. But the pollen record does not indicate any vegetation which would be characteristic of such a climate. It therefore seems likely that the basin has been intermittently subsiding (van der Hammen and Gonzalez, 1964). This alone, however, would not produce the oscillation of Paramo and forest elements, and it seems unlikely that oscillatory tectonic movement could have occurred without leaving massive evidence, which has not been found. The oscillations in the pollen spectra are therefore interpreted as due to actual migration of the vegetation zones up- and down-hill. Presumably this was a response to climatic change, and we immediately ask ourselves whether these were changes of temperature, or hydrologic changes, or both. In the case of the Andes, we do have a good idea of what vegetation is like under very arid conditions, for the Andes in Peru and Bolivia between c.15° and 20°S support just such a vegetation, the Puna, consisting of grasses, small herbs and cacti. We have no modern pollen rain samples from such vegetation, however. It is possible such samples would be similar to those from Paramo. It is therefore impossible to say whether the Gramineae-dominated phases in the diagram represent the moist Paramo or the dry Puna. Also, there could have been considerable reduction in precipitation without the threshold for markedly xerophytic vegetation being reached. The degree of hydrologic change in the climate is therefore uncertain. The changes exhibited in altitudinal zones, however, do clearly suggest temperature changes. These changes have been tentatively correlated (van der Hammen, 1974) with the north-west European glacial sequence.

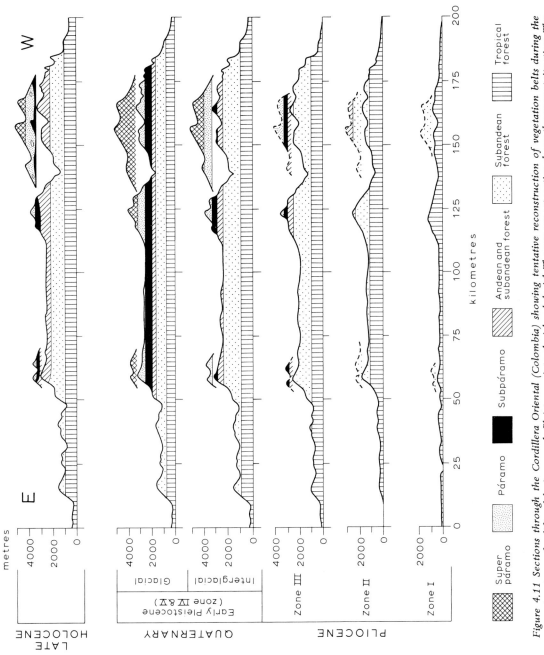
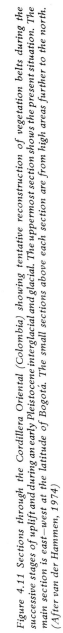

Figure 4.11 Sections through the Cordillera Oriental (Colombia) showing tentative reconstruction of vegetation belts during the successive stages of uplift and during an early Pleistocene interglacial and glacial. The uppermost section shows the present situation. The main section is east–west at the latitude of Bogotá. The small sections above each section are from high areas further to the north. (After van der Hammen, 1974)

64

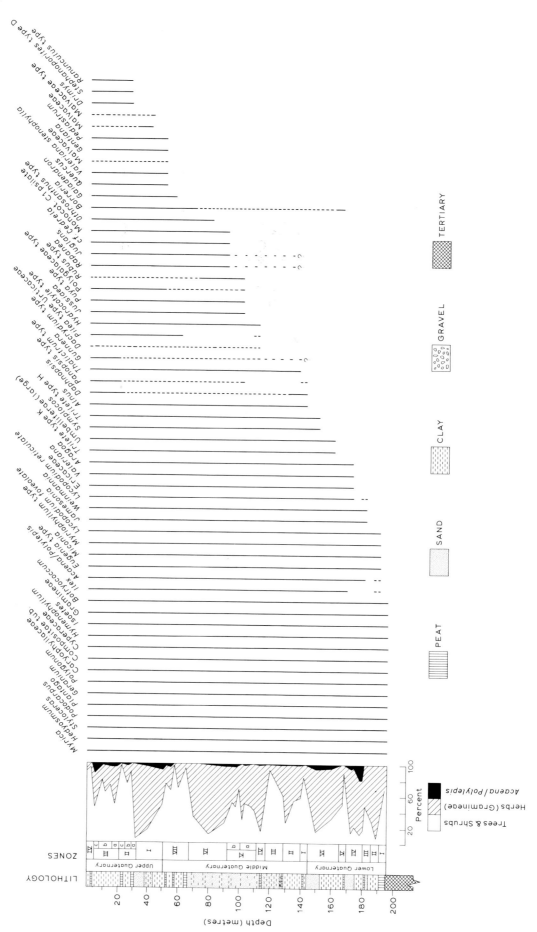

Figure 4.12 Provisional pollen diagram of the 200 m core of Pleistocene lake sediments and peat from the Sabana de Bogotá, Colombia. The progressive immigration of montane elements is demonstrated. (After van der Hammen and Gonzalez, 1964)

The uppermost 33 m of the Sabana de Bogota core was studied in more detail (van der Hammen and Gonzalez, 1960), and appeared to go back well into the ultimate cold phase of the Pleistocene. The last 30 000 years or more of record is also present in a core from the lake of Fuquene on the next high plain to the north of Bogota, and at the same altitude (c.2580 m). The pollen diagram from Fuquene (van Geel and van der Hammen, 1973) is reproduced here in outline as *Figure 4.14*. This diagram is interpreted as follows:

Zone V (c.30 000 to c.25 000 B.P.)

During subzone *V-1 Polylepis* woods dominated on the slopes surrounding the lake. *Quercus* forest probably dominated in the Andean forest belt below the *Polylepis* zone. During subzone *V-2* there was a considerable increase in open vegetation (Gramineae). The *Polylepis* woods declined and in the Andean forest *Quercus* was replaced successively by *Rapanea, Myrica* and *Miconia*, and *Podocarpus*.

The lake level was high at first, and the climate wet. Later (*V-2*) the level may have declined due to a somewhat drier climate. The forest limit during zone *V* may have been about 800-1000 m lower than today, suggesting a cooler climate.

Zone W (c.25 000 to c.13 000 B.P.)

During subzone *W-1* the open vegetation (Gramineae) increased considerably and the lake level was low, suggesting cool dry conditions. In subzone *W-2* a minor decrease of Gramineae and increases of *Polylepis* and *Alnus* may be due to a slightly wetter climate or a slight rise in temperature. Subzones *W-3* and *W-4* represent another minor vegetational oscillation similar to subzones *W-1* and *W-2*, but there are significant changes in forest composition, such as the decline of *Quercus*. Subzone *W-5* (beginning c.20 500 B.P.) saw the complete domination of open vegetation; The forest limit must have been far below the level of the high plateau of Fuquene. The lake level became extremely low, so that the hydrosere extended to the drilling site, as shown by *Myriophyllum*, Cyperaceae and Umbelliferae. Also the sedimentation almost ceased. High values for Compositae and Cruciferae (? *Draba*) suggest Super-paramo vegetation, or at least uppermost Paramo. Everything seems to indicate the climate was exceedingly cold and dry, with the forest limit depressed about 1500 m, equivalent to a lowering in mean annual temperature of c.10°C.

Zone Y (c.13 000 to c.9 500 B.P.)

During subzone *Y-1* (until c.10 800 B.P.), forest again invaded the plateau and its surrounding slopes, probably reaching as high as at the present day. The early importance of *Dodonaea* in this forest is in line with the pioneer qualities of this shrub. The lake filled up again and was warm enough for the alga *Coelastrum reticulatum* to live there, suggesting a climate similar to that of the present. In subzone *Y-2* the forest limit was forced down again, perhaps by 800 m. The lake level fell again, so that a hydrosere of Compositae,

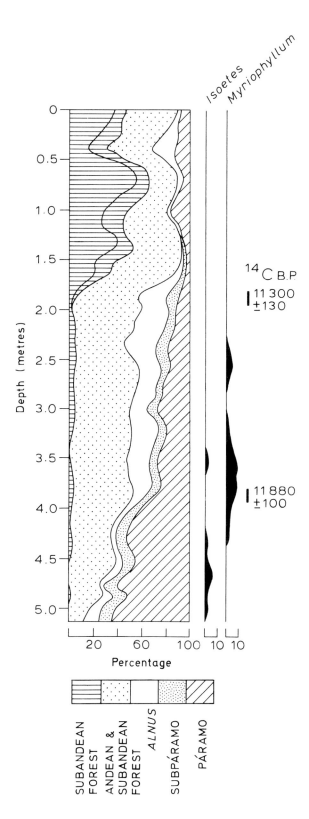

Figure 4.13 Summarised pollen diagram of the Late Glacial and Holocene of Laguna de Pedro Palo, Cordillera Oriental, Colombia (altitude 2000 m). Apparently the forest limit was below 2000 m in the Late Pleistocene. (After van der Hammen, 1974)

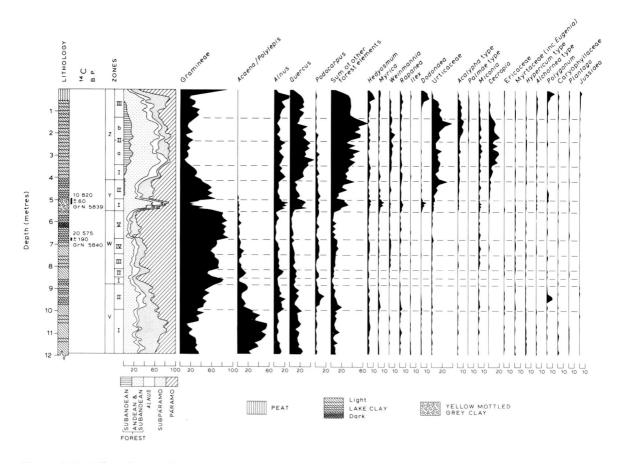

Figure 4.14 Pollen diagram from Laguna de Fuquene, Colombia (altitude 2580 m) covering the Late Pleistocene and Holocene. Only selected pollen types are shown. The values are percentages of a pollen sum including trees, Gramineae and Polylepis/Acaena. (After van Geel and van der Hammen, 1973; van der Hammen, 1974)

Rumex and *Polygonum* could extend over parts of the lake bed. A climate cooler and drier than at present is thus indicated.

Zone Z (9500 to 0 B.P.)

In subzone Z-1 (to c.7500 B.P.) the forest limit rose rapidly, and oak forest reached the area. The lake level rose again, so the climate must have been similar to the present one. During subzone Z-2 (c.7500 to c.3000 B.P.) the climate perhaps became still warmer. Peaks of *Cecropia* and *Acalypha* suggest a climate perhaps $2\,^{\circ}$C warmer than today. After c.3000 B.P. (subzone Z-3) the climate perhaps reached its present state. Increases in the pollen of *Dodonaea* and Gramineae suggest forest clearance by man, indicating the start of Indian agriculture.

There are sufficient pollen diagrams from the last 12 000 years or so of Colombia to examine these for altitudinal and regional parallelism. A site lower than Bogota and Fuquene is provided by Laguna de Pedro Palo at 2000 m (*Figure 4.13*). Although the site is well down in the sub-Andean forest at present, the diagram indicates that the forest limit was below the lake in Glacial or early Late-glacial times. This implies an altitudinal depression of c.1500 m, a remarkably high value. The diagram from Paramo de Palacio (*Figure 4.15*) at 3500 m provides an example from a

higher elevation than Fuquene and Bogota. The site lies at present 200–300 m above the forest limit. The curves are generally similar to those from Fuquene but show, as might be expected, overall lower forest pollen values. Nevertheless some sub-Andean elements occur in the middle of the Holocene, and even in the Late-glacial, suggesting a forest limit at these times considerably above the present one; at any rate above the site.

At an even higher altitude is the site Sierra Nevada del Cocuy. Here polleniferous sediment was located in a moraine-dammed valley at 3890 m. At first sight the pollen diagram (*Figure 4.16*) appears to suggest that in Late-glacial time the area was forested and that this was replaced by Paramo in the Holocene. But the presence of *Polylepis/Acaena* in the lower samples combined with the authors' comment that pollen was sparse in these samples, causes one to look again. Reference to the modern pollen rain diagram (*Figure 4.8*) will show that these features are typical of spectra from the bare high mountain zone (Super-paramo), so that the actual vegetational history is one of bare ground, resulting from cold conditions or actual deglaciation, being colonised during the Holocene by Paramo vegetation.

The most interesting sites to show regional variation are those on opposite sides of the Cordillera Oriental

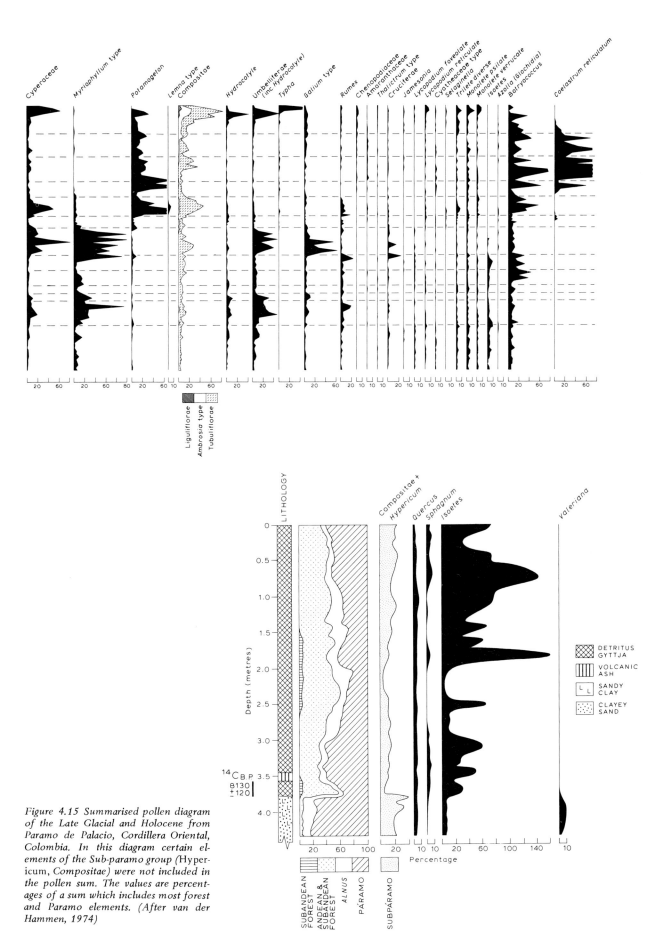

Figure 4.15 Summarised pollen diagram of the Late Glacial and Holocene from Paramo de Palacio, Cordillera Oriental, Colombia. In this diagram certain elements of the Sub-paramo group (Hypericum, Compositae) were not included in the pollen sum. The values are percentages of a sum which includes most forest and Paramo elements. (After van der Hammen, 1974)

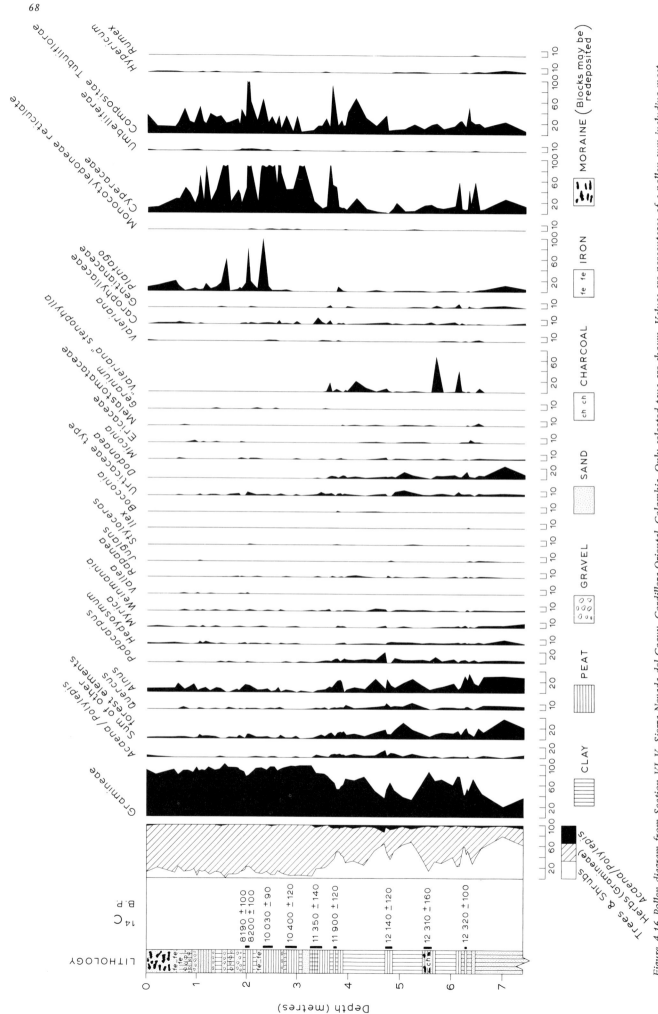

Figure 4.16 Pollen diagram from Section VL-V, Sierra Nevada del Cocuy, Cordillera Oriental, Colombia. Only selected taxa are shown. Values are percentages of a pollen sum including most forest elements plus Gramineae and Acaena/Polylepis. (After Gonzalez et al., 1966)

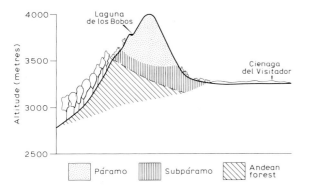

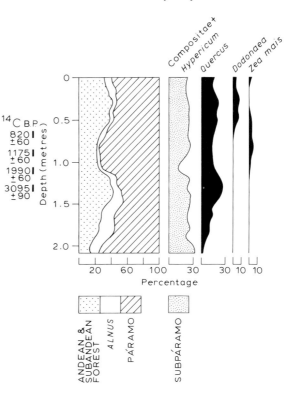

Figure 4.17 *Section through part of the Cordillera Oriental to show the positions of Laguna de los Bobos and Cienaga del Visitador. (After van der Hammen, 1974)*

Figure 4.19 *Summarised pollen diagram from the late Holocene of Laguna de los Bobos (altitude 3800 m) (see* Figure 4.17*). The pollen sum is the same as in* Figure 4.18. *(After van der Hammen, 1974)*

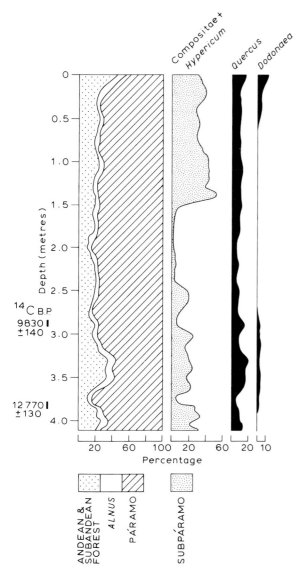

Figure 4.18 *Summarised pollen diagram from the Late Glacial and Holocene of Cienaga del Visitador (altitude 3300 m) (see* Figure 4.17*). The values are percentages of a pollen sum including most forest and Paramo elements but excluding the Sub-paramo elements Compositae and* Hypericum. *(After van der Hammen, 1974)*

near Paramo de Guantiva (*Figure 4.17*). The western slopes have a moist climate and the usual altitudinal zonation, but on the eastern side there is an area about 3300 m altitude which lies in a rain shadow and supports little forest. The pollen diagram from Cienaga del Visitador, in this dry area (*Figure 4.18*) shows a brief development of forest in the Late-glacial, but then a reversion to grassland. This is interpreted to mean a moist Late-glacial interval, followed by an arid Holocene (van der Hammen and Gonzalez, 1965). By contrast, the pollen diagram from Laguna de los Bobos on the wetter western slopes (*Figure 4.19*) although at a much higher altitude, has a far higher percentage of forest elements in the Holocene, which is unfortunately the only time it covers.

The story from Colombia is now beginning to be confirmed and extended by work in Venezuela (Salgado-Labouriau and Schubert, 1976; Salgado-Labouriau, in press).

4.5.2 THE COASTAL AREAS

In coastal areas an additional variable is introduced into the vegetational history. As well as responding to changes in temperature and moisture-availability, vegetation responds to changes in sea level, particularly the eustatic changes brought about by the glacial–interglacial cycle. A core from Ogle Bridge near Georgetown, Guyana (*Figure 4.20*) probably covering the period back to the last interglacial, shows these

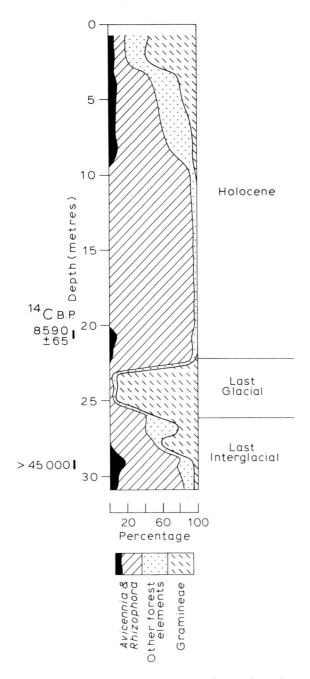

Figure 4.20 Summarised pollen diagram from Ogle Bridge, near Georgetown, Guyana. The evidence indicates savannah during the last glacial time. (After van der Hammen, 1974)

not have occurred if mere build-up of deposits had caused the change in vegetation, rather than a fall in sea level. Presumably the site was well above sea level at this time. There is then a reversion to mangrove. First *Avicennia* swamp passes the site, c.8600 B.P., and subsequently *Rhizophora* dominates. The later Holocene part shows a return to *Avicennia* swamp; presumably this is due to the seral change resulting from rapid silting up during the Holocene, although some change in sea level cannot be ruled out. This diagram is particularly interesting in showing not only the effects of eustatic sea-level changes, but also that the vegetation during (presumably) last glacial time was more likely to have been grass-savannah than rain forest in lowland Guyana. The geographically widespread nature of these changes is indicated by several other cores from Guyana and Surinam (van

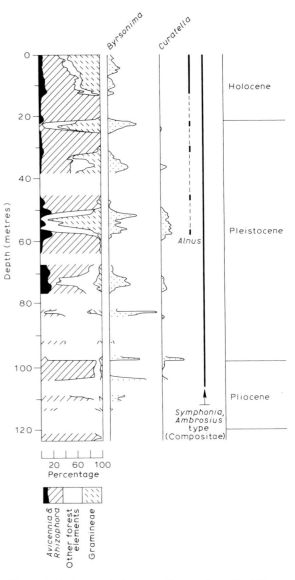

Figure 4.21 Summarised pollen diagram from the Alliance borehole, Surinam. The values are percentages of a pollen sum which includes Avicennia, Rhizophora, *other trees and* Gramineae. *The scale applies to all curves. The diagram shows alternations of savannah and mangrove at the site. (After van der Hammen, 1974)*

changes well. The lower part of the diagram, with abundant *Rhizophora* and some *Avicennia* (a type with low relative export), indicates that the site then lay within the mangrove belt. There follows a peak of swamp forest elements, Palmae, *Symphonia* etc., suggesting the slight retirement of the sea, or the build-up of deposits. Mangrove elements, even the far-spreading *Rhizophora*, disappear completely from the diagram during the next phase when Gramineae pollen dominates, suggesting a grass-savannah. That this was in a dry-land soil rather than in a swamp is indicated by the oxidised-iron stain and other indications of soil formation at this horizon in the core. This would der Hammen, 1963; Wijmstra, 1969, 1971), and the cyclical nature of these changes in the Quaternary is beautifully shown by the 120 m core from the Alliance borehole, Surinam (*Figure 4.21*). Here many of the low-sea-level phases are confirmed as savannahs by the presence of pollen of *Byrsonima* and *Curatella*, characteristic savannah trees. Of course, the presence of savannah at certain times could be the result of edaphic factors, but the probability of an arid or highly seasonal climate in glacial times in the coastal lowlands remains high.

4.5.3 THE INLAND SAVANNAHS

The only long pollen diagram from these areas so far is that from Lake Moriru in the Rupununi savannahs (Wijmstra and van der Hammen, 1966). The diagram (*Figure 4.22*) probably extends well into last glaciation time but is not well dated. In the earliest part of the diagram (c.6–4 m; ?glacial time) there is evidence of a mixture of grass-savannah and savannah woodland (*Byrsonima* and *Curatella*). The succeeding phase

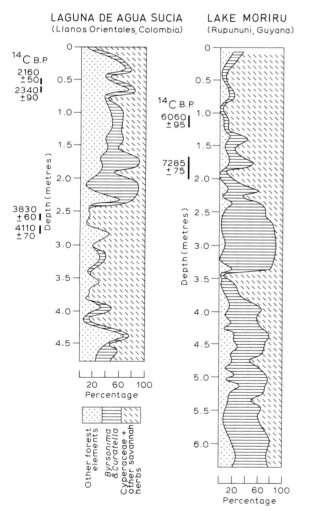

Figure 4.22 Summarised pollen diagrams from the Late Pleistocene and Holocene of two areas at present bearing savannah vegetation in South America. (After van der Hammen, 1974)

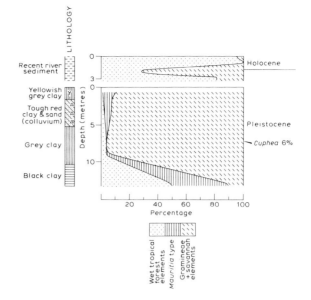

Figure 4.23 Summarised pollen diagrams from Capoeira and Katira, Rondonia, Brazil, showing evidence for former existence of savannah in what is now the lowland rain forest area. (After van der Hammen, 1974)

c.4–3.5 m; ?end of glacial time) suggests a strong increase in grass-savannah. There then follows a pronounced maximum of savannah woodland (c.3.5–2.5 m; ? late glacial). The upper 2.5 m (c.8000 B.P.–present) again suggests dominance by grass-savannah. If we make the assumption that grass-savannah implies a drier climate than savannah woodland, these findings imply, if the estimated dates are correct, a relatively moist climate in Late-glacial times and a very dry climate at the end of the time of the last glaciation.

4.5.4 THE AMAZON BASIN

Considering its great importance as the world's largest area of lowland tropical rain forest, our data from the Amazon basin are ridiculously slight. The only evidence available is a composite pollen diagram from two sections from Rondonia (Brazil, c.9° S) in the southern part of the Amazon basin. The uppermost part (*Figure 4.23*) is from 2 m of recent fluvial sediment, and the lower part from 13 m of Pleistocene valley-fill sediments. The lowest sediment is a humic clay with pollen of lowland rain forest and swamp forest (*Mauritia*-type) elements. In the upper part of the lower section the clays are paler and show intercalation of probable colluvial material. The pollen of this section of the core contains over 90% savannah elements. These are chiefly Gramineae, but include also Cyperaceae, Compositae and *Cuphea*. The latter genus is characteristic of savannahs and the pollen rises to 6% at one point in the core. The section of recent (undated) sediments is dominated by rain forest pollen at the top, as would be expected, but it too shows a phase of dominance by savannah elements at a lower level. These data lead, however inadequately, to the most important suggestion that parts of the Amazon rain forest may have been replaced by savannah during several phases of the Quaternary. Clearly more data from this important area are highly desirable.

4.5.5   CENTRAL AMERICA

Although all the evidence from this area is being treated together as a matter of convenience, in fact there is almost as great a range of environments in Central America as there is in northern South America. From the mountains of Costa Rica at 9 °N, we have a most interesting pollen diagram (Martin, 1964). The site is a bog at 2400 m on an old lake bed, and the diagram (*Figure 4.24*) covers at least the last 36 000 years and possibly the last 110 000 years. The present vegetation is lower montane rain forest dominated by *Quercus*. The diagram was divided by Martin into four zones. The earliest (IV) and penultimate (II) zones are dominated by Gramineae and other Paramo elements. There is a date of c.21 000 B.P. in the middle of zone II. Zones III and I are dominated on the whole by forest pollen. Zone I started slightly before 8000 B.P. These data appear to be in fair agreement, at least where well dated, with those from the Colombian Andes, and do not conflict with the idea of altitudinal shifts of vegetation controlled by temperature changes. There is no very clear evidence of changes which could be interpreted as requiring hydrologic climatic shifts.

Central America also provides us with some evidence from the lowlands. Deep cores from Panama were investigated by Bartlett and Barghoorn (1973). The sediments originated from deposition in a coastal swamp. Because of the strong likelihood of finding river transported pollen in these sediments, they are not very useful for indicating dry land vegetational history. The pollen includes, however, a few montane types such as *Iriartea* which may indicate a cooler climate prior to 7300 B.P. The palm *Iriartea* today occurs only above 1200 m. The highest hills in the area are only 300 m high, so even if the pollen is transported, a depression of range of 900 m is suggested. These analyses also provide evidence of mangrove developments in post-glacial time similar to those in Guyana, and of cultivated plants (*see* Chapter 7).

There are several other short pollen diagrams from Central America (Tsukada in Cowgill *et al.*, 1966; Tsukada and Deevey, 1967), but these are chiefly of interest in connection with the effects of man and will be mentioned in Chapter 7.

4.5.6   THE GALAPAGOS ISLANDS

These biogeographically classical islands, c.1000 km east of South America, possess a semi-desert vegetation

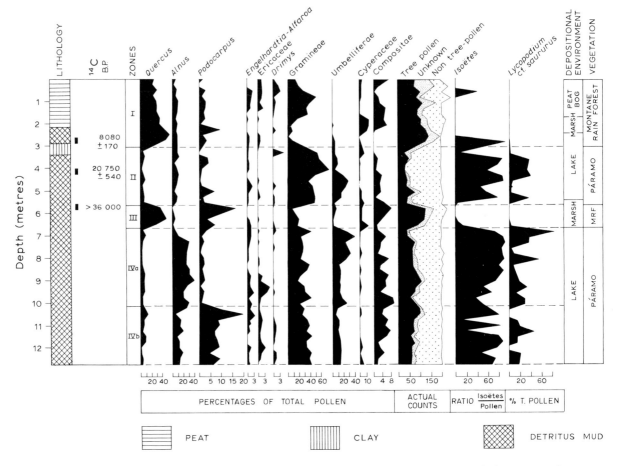

Figure 4.24 Pollen diagram from Vicente Lachner Bog, Costa Rica, altitude 2300 m. Values are percentages of total pollen except for Isoetes *microspores which are ratios of* Isoetes *to total pollen in a 200 grain count of both. (After Martin, 1964)*

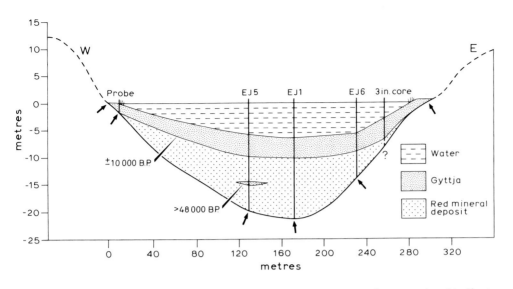

Figure 4.25 Diagrammatic section of El Junco Crater, Galapagos. Vertical exaggeration 5X. Short arrows mark the points at which the rock floor of the crater was identified. The gyttja represents wet periods and the red mineral deposit dry periods. (After Colinvaux, 1972)

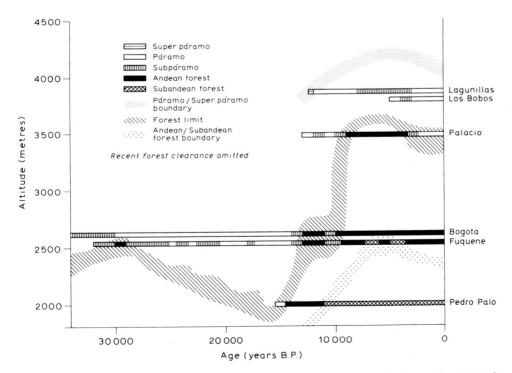

Figure 4.26 Summary diagram of Late Quaternary vegetational changes in the Colombian Andes. (Original)

and climate at present. Nevertheless, the climate is moist enough to permit the survival of a few crater lakes, the deposits of which have now been sampled (Colinvaux, 1968). The most interesting section is that from El Junco (Colinvaux, 1972), shown in *Figure 4.25.* The red mineral deposit probably indicates a dry crater, while the gyttja (organic mud) clearly implies existence of the lake. The latter therefore appears to have existed for a while before 48 000 B.P., then to have dried up and to have re-formed about 10 000 B.P. The two gyttja deposits contain abundant pollen and spores, but there are striking differences between the lower and upper floras. The lower one contains *Myriophyllum* and *Azolla filiculoides*, but no *A. microphylla;* the upper lacks *Myriophyllum* and possesses *Azolla microphylla*, but no *A. filiculoides*. The implication is that the pre-48 000 B.P. flora became completely extinct in the Galapagos during the very dry phase which dried up the lake. Since the lake re-formed, a different species of *Azolla* has immigrated. In view of the isolation of the islands such an explanation does not seem unreasonable; both *Azolla* species occur in South America.

The general correlation of this climatic history with that in other equatorial areas is rather striking. Several different explanations have been advanced for the occurrence of the very dry phase. Colinvaux (1972) suggested it was caused by the inter-tropical convergence zone remaining north of the equator, while Newell (1973) thought the same feature might have been shifted south of the equator. Houvenaghel (1974) explained it, however, in terms of reduction in sea water temperature, related to changes in the intensity of the southern trade wind, and this explanation was adopted by Simpson (1975).

4.6   CONCLUSIONS

4.6.1   VEGETATION

The northern Andes came into existence by uplift in Late Pliocene time. They became gradually vegetated by a mixture of local elements derived by evolution from the lowland rain forest flora, and of migratory elements which arrived from the northern and southern temperate regions. This process continued throughout the Pleistocene and is probably still going on.

The Pleistocene shows alternations of vegetation through time at any one site. Interpretations of dated sites at different altitude show that these changes can be interpreted as a migration of vegetational zones up and down the Andes. These changes are summarised in *Figure 4.26.* Thus during the period 20 000–14 000 B.P. the forest limit appears to have stood, at least on one occasion, as low as 2000 m. Present forest limits are around 3200–3500 m or, under extremely arid conditions, 3000 m (van der Hammen, 1974). Thus a depression of the forest limit by at least 1000 m is indicated. There is so far little direct evidence as to

whether other vegetational zone boundaries were depressed by the same amount, but van der Hammen (1974) assumes a depression of only 500 m for the upper boundary of the lowland forest in his tentative reconstruction (*Figure 4.27*) of the former vegetational zones.

The lowland vegetation was, however, altered in other ways during the Pleistocene. The evidence from coastal sections suggest that there is a long history of savannah in those areas. The Rupununi savannahs themselves are now shown to have existed, at least as savannah woodland, during the Pleistocene. In addition, there is slight evidence that even in the Amazon basin itself, the 'heartland' of the rain forest, savannah may have existed during the Pleistocene. Although much more evidence will be required before a firm conclusion can be reached, it now seems likely that savannahs were much more widespread in the Pleistocene than was until recently believed. Furthermore, the coastal evidence suggests that the times of the existence of savannah were essentially the times of eustatically lowered sea level, i.e. the times of world-wide glaciation. The coastal evidence also indicates that mangrove vegetation has had to migrate back and forth to keep pace with sea level change.

4.6.2   CLIMATE

The depression of vegetation belts during the Late Pleistocene argues, if one assumes the usual relationship between vegetation and temperature, for a climate cooler than at present. A depression of 1000 m, using the usual values for lapse rate, would imply a mean annual temperature lower by some 6–7 °C in the Andes (van der Hammen, 1974). This is a surprisingly high figure, since the lowering of the temperature of the surface waters of the Caribbean at the same time was possibly only 2–3 °C (van der Hammen, 1974 – evidence obtained from foraminifera and $^{16}O/^{18}O$ data). It is possible that lapse rates were different at the time.

The evidence regarding possible changes in hydrology is also fairly clear. In the Andes some increase or decrease in precipitation might have occurred without very distinct influence on the vegetation. There is evidence from Cienaga del Visitador for a wetter climate some time between 12 800 and 9800 B.P., and this is supported by evidence of high lake level at Fuquene. The same lake appears to have been at a low level during the period 20 000–14 000 B.P.

In the lowlands the evidence is more scanty but what exists is less equivocal. It appears that drier climates may have prevailed in the past. Precisely when this happened is not certain in all areas, but the times of arid climate possibly coincided with low sea level, i.e. glacial times. On the other hand, the lowland evidence has little to tell us about temperature changes: that is to say, there is little in the lowland vegetational evidence which requires there to have been temperature changes there.

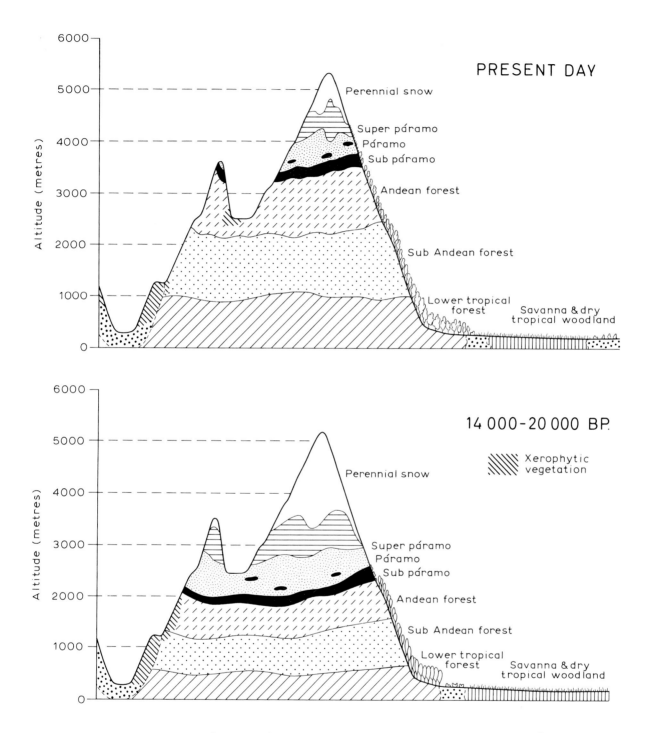

Figure 4.27 Tentative reconstruction of vegetational zones in the Eastern Cordillera of the Andes, near Bogotá, during the Last Glacial maximum compared with present vegetation. (After van der Hammen, 1974)

4.6.3 BIOGEOGRAPHICAL PROBLEMS

The vegetational history allows us to suggest solutions to several biogeographical problems. The mountain problems of disjunction, endemism and vicarism may be viewed in the light of the evidence for repeated depression of the vegetational zone boundaries during the Pleistocene. In particular, the former connection of many Paramo and Super-paramo 'islands', followed by their isolation as the forest limit rose, provides a good explanation for disjunct distributions. A reconstruction of Paramo areas about 16 000 B.P. (*Figure 4.28*) suggests that most Paramo islands were then connected, or at least the gaps between them were much reduced, and earlier cold phases may have been even more severe. The same oscillations can also explain vicarism, which is the expected result of isolation after former geographical connection.

The problems of savannahs must await more data before a firm conclusion about them can be given, but there is already sufficient evidence to say that savannahs have existed in some form for a very long time in South America and are therefore a 'natural' plant formation. It seems likely that at times in the past savannahs have been very much more widespread than at present. This is in accordance with Eden's (1974) conclusions that the similarity of floras of

relict patches of savannah demanded former contact between them. It also does not conflict with the conclusion of Labouriau (1966) that the Brazilian Cerrados must have co-existed with forest for a long period. This conclusion was based on the fact that there are at least 35 pairs of vicarious species of plants between cerrado and forest, which probably required a long time to evolve.

If the scanty Amazonian evidence of former savannah occurrence can be confirmed, it may well transpire that the lowland rain forest was indeed divided into separate blocks by the extension of savannahs in the Pleistocene. This would not only satisfy the requirements of the zoogeographers (Haffer, 1969; Vanzolini and Williams, 1970; Simpson, 1971; Brown, 1972) but also possibly help to explain the diversity of the lowland forest (Prance, 1973). If, as is postulated for birds and reptiles, the species in each rain forest block became genetically isolated from those in other blocks, the subsequent amalgamation of the blocks would lead to clusters of closely related species — a familiar rain forest phenomenon. It is not suggested this is a complete answer to the problem of floristic diversity in the lowland rain forest of South America, but it may be a contributory cause. It can however, be objected to on the grounds that similar arid phases in Africa were used as an explanation of reduction in diversity, rather than its increase.

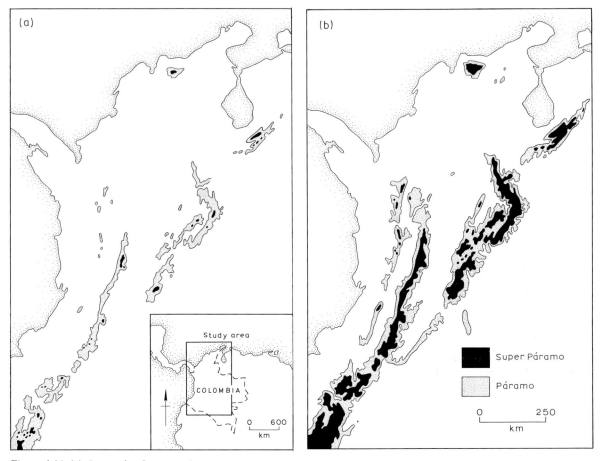

Figure 4.28 (a) Geographical extent of present vegetation zones in the northern Andes compared with (b) the likely extent of Paramo and Super-paramo during the Last Glacial maximum. The area indicated as Super-paramo on the right was partly occupied by glaciers. (After van der Hammen, 1974)

5
The Quaternary Vegetation of Equatorial Indo-Malesia

5.1 INTRODUCTION

In some ways this area has undergone more geomorphological change during the Quaternary than Africa and South America, and it is difficult to reconstruct accurately the condition it must have been in at the start of the Era. The major changes during the Quaternary have been of three chief kinds: volcanic, tectonic and eustatic.

Malesia is one of the most volcanically active regions of the world. Most of the major peaks of Java and Sumatra, and some of those of New Guinea, are Quaternary volcanoes, active or extinct, and the amount of the surface composed of Quaternary volcanics is considerable.

Most of the non-volcanic peaks in Borneo and New Guinea have resulted from uplift either partially or wholly during the Quaternary. Thus Mt Kinabalu in northern Borneo is a granite batholith with potassium-argon ages as low as 1.5 M years (Jacobson, 1970) and parts of the north coast of New Guinea are still rising very rapidly (Chappell, 1973). Subsidence has also occurred, a striking example being the Semangko rift valley which runs the length of Sumatra (van Bemmelen, 1949). Some geomorphological features are, of course, much more ancient. The West Malaysian mountains, for example, are believed to date from the Mesozoic (Richardson, 1947).

A eustatic fall in sea level of 100 m would have influenced Malesia far more than Africa or South America. Most of the shallow South China Sea and all of the Torres Strait would have been dry land. That a

low sea level actually occurred on the Sunda shelf is proved by submarine peat deposits (Aleva, 1973; Biswas, 1973). Kunkar nodules on the Sahul shelf point to arid dry land conditions there during at least part of the last glaciation (van Andel *et al.*, 1967). The drainage channels of these areas when they were exposed have been mapped (*see Figure 5.1*). Despite this great increase in extent of the Sunda shelf from Asia and the Sahul shelf from Australia, there are no geological grounds for believing that continuous land bridges existed between the two at any time in the past, although the sea barriers may have been very narrow at times. It was also formerly thought that there was evidence in Malesia of Quaternary sea levels much higher than at present, but it has been shown by Haile (1971) that this is most unlikely, apart from the c.6 m Holocene level found on the east coast of West Malaysia (Fitch, 1952; Nossin, 1962) and in other areas.

5.2 PRESENT VEGETATION

Before interference by man, by far the greatest part of the lowland in this whole region was occupied by rain forest (Whitmore, 1975), and there are vast surviving areas of this vegetation in Borneo, Sumatra, New Guinea and elsewhere (*Figure 1.5*). The rain forest flora is similar throughout Malesia (i.e. the political states of Malaysia, Indonesia, the Philippines, and Papua New Guinea), but sufficiently different in South India and Sri Lanka for these to be placed in a

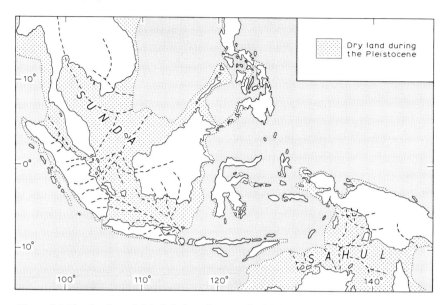

Figure 5.1 *The Sunda and Sahul shelves showing the likely maximum extent (dotted lines) of dry land during the Pleistocene. The major drainage channels then existing are indicated by broken lines (After Verstappen, 1975)*

different botanical region by Good (1947). The flora of the equatorial Pacific isles can best be regarded as a depauperate Malesian flora (van Balgooy, 1971).

There are significant areas of savannah and related vegetation types (such as semi-evergreen forests) in South India, Sri Lanka, Burma, Thailand, East Java, the Lesser Sunda Isles and southern New Guinea, but these will not be considered further as we have very little direct evidence of their history.

Mountain vegetation is abundant and diverse in the region (Whitmore, 1975). The montane zonation is very variable, so mention of it here will be restricted to those areas which will feature later in the chapter. In New Guinea it is extremely difficult to generalise about the zonation for every mountain range is a law unto itself. For instance the uppermost woody vegetation on part of the Central Range in Irian Jaya includes *Casuarina* sp. as an important element (Lam, 1945), yet this species is totally absent on the other mountains of New Guinea. Most of the palaeo-ecological work has been done in Papua New Guinea, and much of it on the highest peak there, Mt Wilhelm (4510 m). The general zonation in this area may be summarised in *Figure 5.2*, although it must be emphasised that there is a continuum, particularly in the forest, rather than distinct zones.

The second important field area is Central Sumatra. Here the vegetational zonation differs depending on the age of the mountain which is being studied. Long-quiescent or extinct volcanoes and non-volcanic peaks have an apparently stable vegetation which conforms approximately to the scheme formulated by van Steenis (1934–36). The lowland rain forest gives way at about 1000 m to a lower montane rain forest which is very mixed but leans towards single family dominance by the Fagaceae. The genera here, however (*Quercus, Lithocarpus* and *Castanopsis*), do not include the *Nothofagus* of New Guinea. At around

1800 m there is a changeover to forest in which *Podocarpus* spp. are abundant, although these also occur at lower altitudes. Above about 2800 m there are dwarf forests rich in Ericaceae and *Myrica javanica*. Open, apparently stable, vegetation occurs in North Sumatra on the summit of the granite Gunong Leuser at c.3300 m, and this is a rhododendron scrub in which *Primula imperialis, Gentiana quadrifaria* etc. emphasise the temperate relationships (van Steenis, 1938).

The vegetation of volcanoes varies widely depending on their age, size, shape, type of effluvium etc., but we will discuss here only the intermittently active Mt Kerinci (3800 m), the highest peak in Sumatra. This is an acidic cone, producing showers of scoria

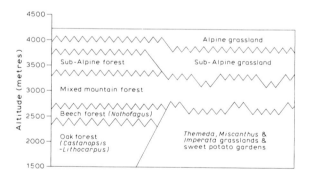

Figure 5.2 *A scheme of zonation for vegetation in the New Guinea mountains. (After Walker 1970a)*

and ash. The lower montane forest on the lower slopes is indeed fagaceous – and coniferous – dominated, but floristically it is much poorer than that at the same altitude on the nearby Mt Tujuh (Jacobs, 1958). At about 2400–2900 m the forest is interrupted by a discontinuous ring of patches of non-forest vegetation dominated by the fern *Gleichenia*. The mechanism of origin and maintenance of these patches is not clear; they could result from the burning of the forest by man, or perhaps from a previous large eruption. Above 2900 m is an Ericaceae/*Myrica javanica* forest, rather dwarf, which gives way about 3200 m to *Vaccinium* scrub growing on the freshly deposited dry scoria. This scrub becomes more and more dwarf at higher elevations, and the mountain is more or less bare above 3400 m to the summit on the crater rim at 3800 m. Most other Sumatran volcanoes have a similar occurrence of *Vaccinium* scrub on the newly deposited ash at various altitudes. *Myrica javanica* can also contribute to these vulcanoseres, and in North Sumatra *Pinus merkusii* forms dense stands, possibly on old volcanic effluvia (van Steenis, 1972; Whitmore, 1975).

Since it is intended in this chapter to mention some historical evidence from 17 ° S, in Queensland, Australia, it is appropriate to consider briefly the vegetation of northern Queensland. Here a rain forest similar floristically to the southern New Guinea forests occurs along the coast (Webb and Tracey, 1972). The rain forest takes different forms depending on altitude, rainfall and edaphic factors (Webb, 1959, 1968). Inland, on the Atherton Tableland, rain forest gives way to dry sclerophyll vegetation dominated by species of *Eucalyptus* and *Acacia*, with a ground flora of Gramineae. The boundary between the two is often marked by a narrow belt of transitional wet sclerophyll forest with *Casuarina*. The rain forest boundary is by no means a straight one, and there are frequent outliers, but the overall trend from rain forest in the east to dry sclerophyll in the west is very clear indeed (*Figure 5.3*).

5.3 BIOGEOGRAPHICAL PROBLEMS

One of the chief problems which has taxed the minds of biogeographers in relation to Malesia is the origin of the mountain floras. The individual peaks in Java, for example, have rather similar floras, especially above the forest limit (van Steenis, 1934—36), and the species involved may show remarkable disjunctions. One of the most extreme disjunctions in Malesia is that of *Drapetes ericoides* which occurs, within Malesia, in New Guinea and on Mt Kinabalu in North Borneo (*Figure 5.4*), a disjunction of some 2000 km. Other examples are *Euphrasia* (*Figure 1.7*), *Primula imperialis* (= *P. prolifera*), and *Haloragis micrantha* (van Steenis, 1962a). Some of these (*Euphrasia*, *Primula*) are of north temperate affinities; others are related to south temperate floras, so that van Steenis (1934—36) was able to postulate the existence of three separate migration tracks for montane taxa (*Figure 5.5*). The Sumatran track was believed to have been the route for taxa from the Himalayan region (e.g. *Primula imperialis*). The Taiwan—Luzon track is postulated for taxa from the north (e.g.

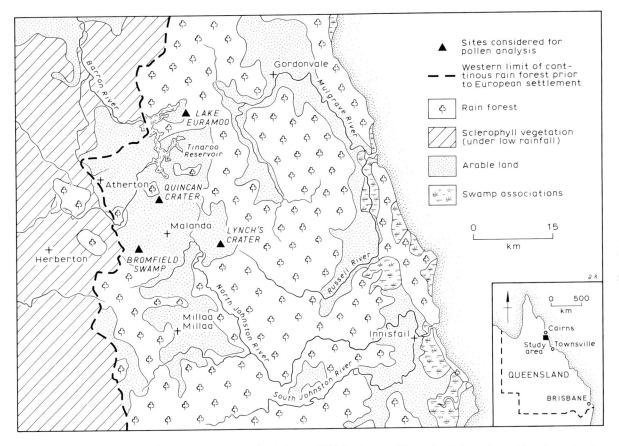

Figure 5.3 The distribution of vegetation types in the Atherton Tableland area of Queensland, Australia, and locations of pollen sites. (After Kershaw, 1973a)

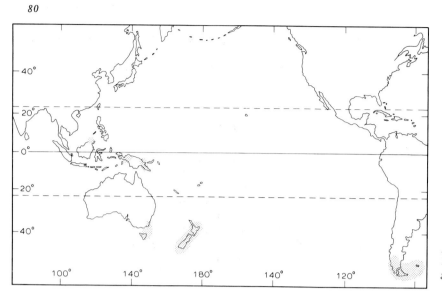

Figure 5.4 The distribution of the genus Drapetes (including Kelleria), showing great disjunction. (After van Steenis, 1964)

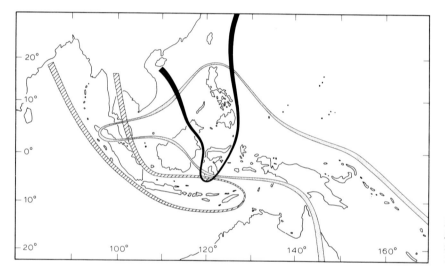

Figure 5.5 The three main tracks of Malesian mountain plant species from their source areas. (After van Steenis, 1964)

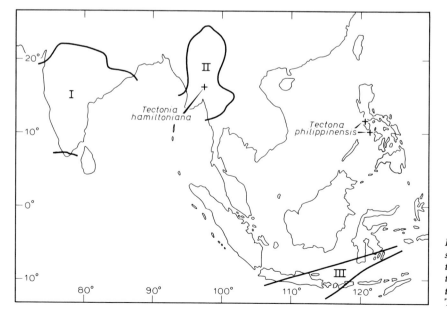

Figure 5.6 The distribution of Tectona spp. illustrating the disjunction between the dry zone of eastern Indonesia and that of Burma–Thailand. I, II and III: the three regions of occurrence of Tectona grandis. (After Hedegart, 1976)

Lilium) and the New Guinean track for taxa of south temperate affinities (e.g. *Drapetes, Astelia*).

It was originally envisaged by van Steenis that migration had occurred along formerly continuous mountain ranges. Now that geology has failed to confirm the existence of such ranges, the migration tracks must be thought of as routes for long distance dispersal (Smith, 1974), using mountains as stepping stones. Clearly Quaternary climatic change, bringing lower hills within the altitudinal range of tropicalpine species, is likely to have been of great significance. It is therefore most desirable that direct evidence of the use of these routes should be sought.

A disjunction exists between the lowland flora of the dry zone of East Java and the Lesser Sunda Isles and that of Thailand (Ashton, 1972; Verstappen, 1975). The classic example here is Teak (*Tectona grandis*) which is the native dominant of semi-evergreen forest in both areas (*Figure 5.6*). On the other hand, many of the species from the driest areas of Thailand—Burma do not occur in similarly dry areas in the Lesser Sunda Isles. The same is true of mammals; the true horses (*Equus*) and a number of antelopes and other bovids, all of which are adapted to savannah environments, are absent from the dry zone of East Java and the Lesser Sunda Isles. Given formerly continuous land connections, these differences suggest an ecological barrier, of which lowland rain forest is the most likely to have occurred (Medway, 1972). Clearly it would be desirable to have palynological information on the history of the teak forest and related vegetation types, to check whether or not a corridor of savannah existed during the Quaternary for the migration of species between Thailand and East Java, as proposed by van Steenis and Schippers—Lammertse (1965).

A third problem is provided by Wallace's line (*Figure 5.7*). This most famous of all biogeographical boundaries marks the division between the Asiatic fauna of placental mammals and the Australasian fauna of marsupials. The original line drawn by Wallace ran between the islands of Bali and Lombok; later, Weber redrew the line to the east of the Celebes, and it is now generally recognised that the Celebes and associated islands should be regarded as a transitional area which has been called Wallacea (*Figure 5.7*). Actually, the Wallacean fauna is heterogeneous (Mayr, 1944). A few, but very few, groups of placental mammals have made the complete crossing to Australia—New Guinea: bats (which could have flown), rodents (perhaps transported accidentally), dogs (probably brought by man). The line is also a remarkably good demarcation for avifaunas. Why is not Wallace's line a botanical demarcation? The best botanical dividing line comes in southern New Guinea (Walker, 1972). Few mountain plants are apparently restricted by Wallace's line and there is a rather gradual transition of montane floras across Malesia (van Steenis, 1934—36). The fossil evidence suggests that the origin of this situation may not be a Quaternary problem, for the existing floras and faunas appear (from the fragmentary evidence) to have been roughly as they are now for some time back into the Tertiary (*see* Chapter 2), but the question

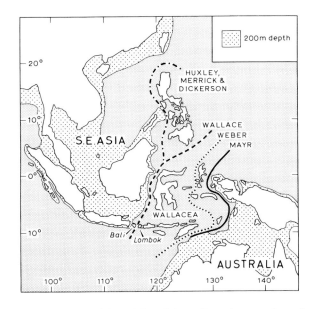

Figure 5.7 South-East Asia and Australia to show continental limits (200 m depth) and biogeographical boundaries. (Mainly after Mayr, 1944; Keast, 1959; Carr, 1972)

why, say, the avifaunas did not mingle more may be a matter of ecological rather than ocean barriers, and may therefore be susceptible to answer by Quaternarists.

5.4 MODERN POLLEN RAIN STUDIES

In Malesia modern pollen rain studies have been carried out by several workers, although there has still been no study of a thoroughness comparable with that of Hamilton (1972) in Africa. In the New Guinea Highlands, Powell (1970) examined moss polsters and similar samples from all the main vegetation types. A brief summary of her data, along with additional spectra (Flenley, 1967) is presented in *Figure 5.8*. The results shown are quite encouraging as regards the possibility of interpreting fossil diagrams. Spectra from areas designated 'non-forest' (chiefly gardens, regrowth and herbaceous swamps) are generally dominated by pollen of Gramineae, or of pioneer trees such as *Trema, Macaranga* or the mostly planted *Casuarina*. Forest site samples, on the other hand, are dominated by forest pollen. Within the forest, 'oak' forest samples tend to be dominated by the pollen of the 'oaks' *Lithocarpus* and *Castanopsis*, 'beech' forest by *Nothofagus* pollen. The highest altitude mixed forests yielded samples in which many forest pollen types occurred (e.g. *Podocarpus* and *Papuacedrus*, not shown in the diagram), but in which *Quintinia* pollen, not common at other altitudes, was regularly present. Things are different in the tropicalpine areas above the forest limit. Gramineae are again abundant here, presumably as a result of local production by the tussock grassland. Pollen of forest and non-forest trees is however carried uphill abundantly and may reach values of 20% for individual taxa, particularly where local vegetation is sparse, as at the summit of Mt Wilhelm (4510 m). The situation here is closely analogous with

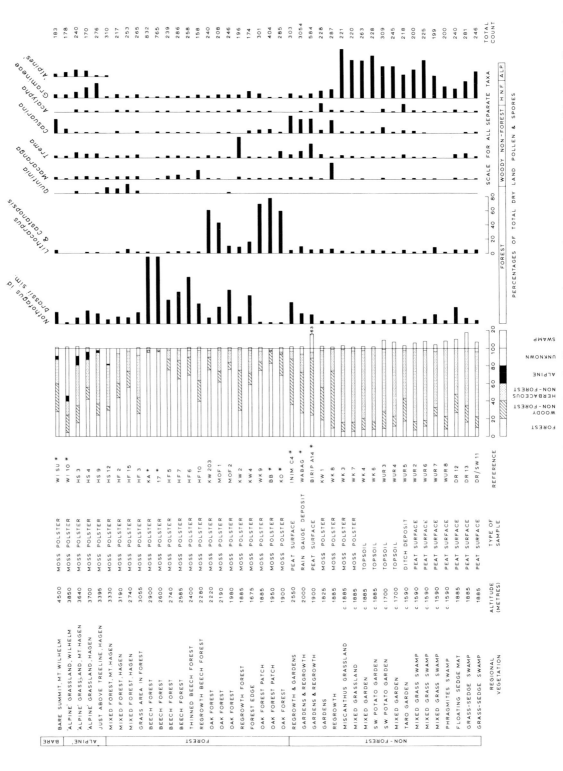

Figure 5.8 Modern pollen rain from the New Guinea Highlands. Only selected pollen types are shown. Values are percentages of total dry land pollen and spores. All data are from Powell (1970) except those marked *which are from Flenley (1967). Powell samples include Ranunculus in 'alpines' whereas Flenley samples include this in 'herbaceous non-forest'. (After Flenley, 1973)

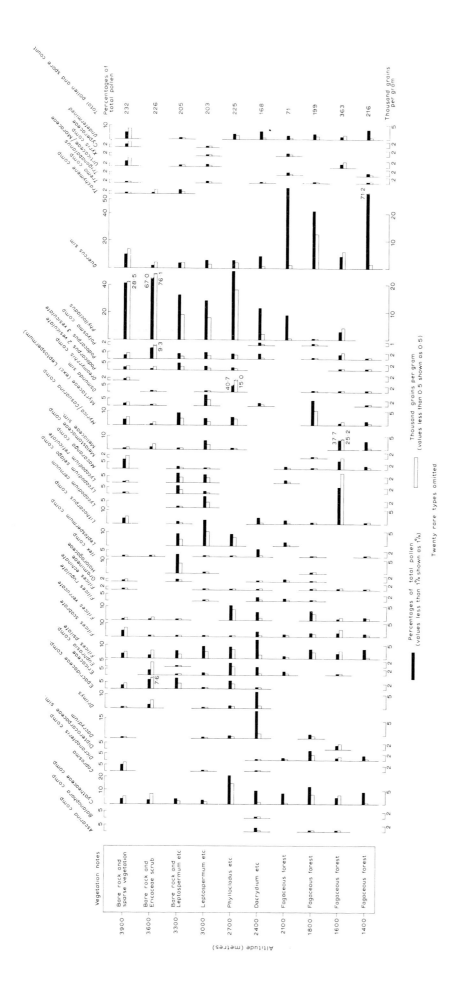

Figure 5.9 Modern pollen rain from Mt Kinabalu, Borneo. The carriage of forest pollen uphill is very well demonstrated. (After Flenley, 1973)

84

KERINCI AREA

Modern pollen rain at primary forest sites & relation to primary vegetation

Pollen sum total primary arboreal pollen

Arboreal taxa only

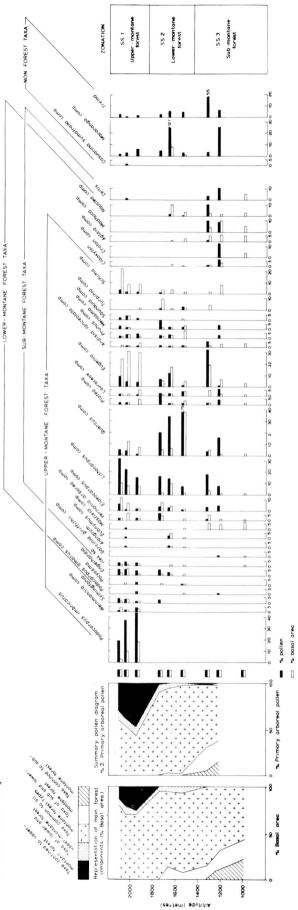

Figure 5.10 Modern pollen rain from the Kerinci area of Sumatra and its relation to present vegetation. (After Morley, 1976 and Flenley et al., 1976)

that in the Super-paramo in South America. Fortunately these tropicalpine spectra may be identified by the presence of pollen taxa referable to tropicalpine species. The most abundant of these, in pollen terms, is usually *Astelia papuana,* a small liliaceous herb only found in herbfields and bogs above 3050 m (Flenley, 1972). As regards pollen of the upper forests and tropicalpine areas, Hope (1976a) has refined and furthered these conclusions, with particular reference to Mt Wilhelm. As a result of the analysis of 80 samples collected from 22 sites, using Tauber traps (Tauber, 1967) Hope has concluded:

1. Pollen is carried upslope effectively, but not downslope.
2. About 70% of the pollen reaches traps as a rainout component.
3. The regional pollen rain is about 800–1000 pollen grains per cm^2 per annum, equivalent to figures reported from temperate rather than tropical lowland areas (Kershaw and Hyland, 1975).
4. Pollen deposition within plant communities (local plus regional deposition) ranges from 1000–7000 pollen grains per cm^2 per annum.
5. All communities sampled have distinctive pollen spectra reflecting the local deposition and its effect on the regional spectrum. For example communities with large local deposition show a depression of the percentages of the regional elements relative to spectra from communities with low local production.
6. Pollen trapping shows that regional elements are deposited in similar amounts at all sites. Filtration (*sensu* Tauber, 1967) is apparently not significant, possibly due to the high rainfall.

Studies in other montane regions have suggested similar findings. A particularly clear example is provided by the outline data from Mt Kinabalu (*Figure 5.9*) (Flenley, 1973). Kinabalu is peculiar in that the usual tropicalpine grassland is almost missing: forest and scrub give way at c.3000 m to bare granite. This is reflected in the pollen rain precisely as one would expect; forest pollen types (especially *Phyllocladus*) become dominant in samples from the bare granite, although tropicalpine indicators such as *Trachymene* and *Oreomyrrhis* do exist. Below the forest limit, pollen spectra reflect reasonably faithfully the vegetation in which they occur e.g. conifer pollen dominates in spectra from conifer forests and pollen of *Quercus* sim. and *Lithocarpus* comp. in samples taken from forests where Fagaceae are generally dominant.

Turning now to somewhat lower altitudes, Morley (1976) has shown a close relationship between forest composition and pollen spectra in Sumatra between c.1000 and 2000 m (*Figure 5.10*). Here a separation was clear between spectra from his upper montane forest, above 1800 m (dominated by *Dacrycarpus imbricatus* pollen), those from his lower montane forest, 1400–1800 m, (dominated by *Quercus* comp. pollen) and those from his sub-montane forest,

1000–1400 m, dominated by pollen of various taxa including *Celtis* (which is absent in almost all spectra above this). It should be pointed out that these forest categories differ from those defined elsewhere for Malesia (e.g. Whitmore, 1975).

If modern pollen rain studies in the Malesian highlands are at an early stage of development, those at lower altitudes must be regarded as almost pre-natal. In the lowlands the three great problems of pollen diversity, relative rarity of anemophily and relative windlessness are believed to reach their apotheosis. An additional problem is the fact that moss polsters are rare in the lowland forest.

Surface samples from lowland swamp forest in Borneo were analysed by Muller (1965) who found that lateral movement of pollen was indeed very restricted, so that samples tended to be dominated by the nearby trees.

In West Malaysia pollen spectra from three Oldfield traps have been analysed by Morley (in Flenley, 1973) (*Figures 5.11 and 5.12*). Two spectra come from the floor of the rain forest and the third from a platform 43 m up in a tree of *Anisoptera laevis*. The catch of 800–2020 pollen grains and spores per cm^2 per annum was lower than those for temperate regions (1000–6000 grains per cm^2 per annum), but not markedly so. The composition of the spectra from these traps is also extremely interesting, although the paucity of data makes conclusions impossible. The quantity of *pollen* caught by the canopy trap was intermediate between the pollen catches of the two ground traps. On the other hand, the quantity of *fern spores* in the ground traps was at least twice that in the canopy trap. On a percentage basis (*Figure 5.12*) these facts become even clearer. As percentages of total pollen, the ground traps each had about 50% psilate fern spores; the canopy trap had 10%. For *Asplenium* comp. the figures are 15% and 3% and for *Nephrolepis* comp. 15% and 6%. But the really interesting thing about these lowland forest samples is the sheer diversity of the catch. The mean number of types recognised (many still unidentified and capable of subdivision) was 60 in a count of 925 in the ground traps and 62 in a count of 779 in the canopy trap. In a detritus sample from mixed forest at 213 m near the foot of the Cameron Highlands road, West Malaysia, 60 pollen and spore types were recognised in a count of 621 (including extra traverses examined for presence only). In this sample 91% of the pollen (excluding spores) was of one type (*Eugenia* comp.) and there was a tree of *Eugenia* sp. directly above, yet still the fantastic diversity was maintained.

Other workers have drawn attention to this diversity in relation to recent fossil material. Polak (1933) in the cores she studied from Java and Sumatra found a very large number of types (although few were identified). Muller (1965) lists 51 types found in peat in Sarawak (not all in one sample) – and all these are at the generic or family level, so that a large number of species could be represented, although the forest in the peat swamps is relatively undiverse (Anderson, 1963), compared with dry-land rain forests.

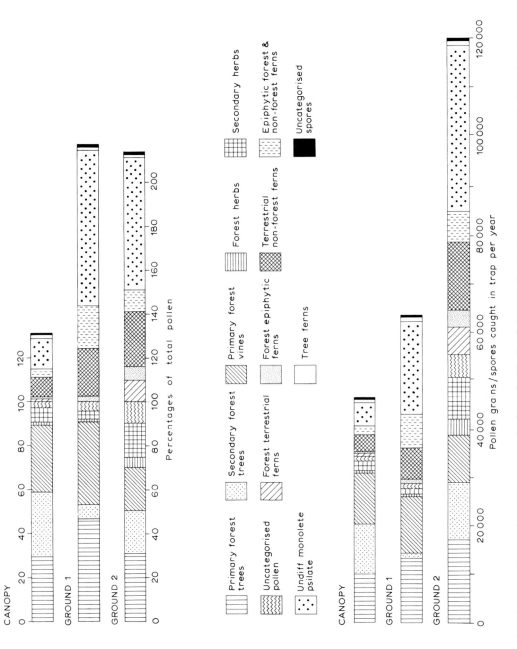

Figure 5.11 Summary diagram for pollen trap samples from the Ulu Gombak Virgin Jungle Reserve, Selangor, W. Malaysia. (After Flenley, 1973)

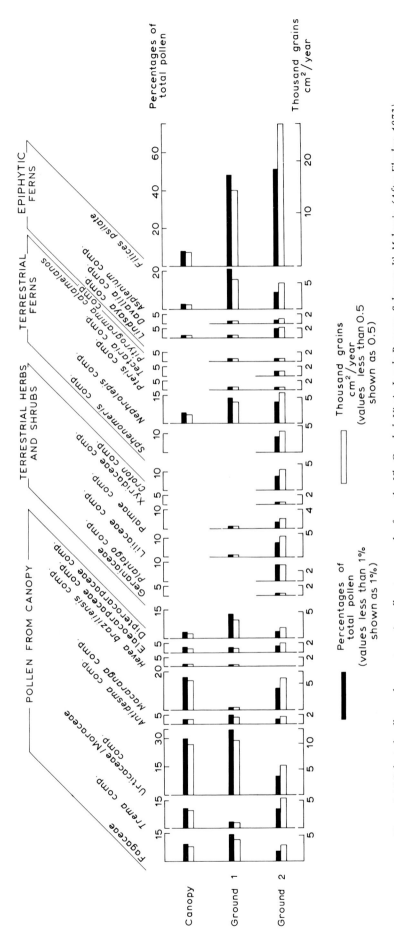

Figure 5.12 Selected pollen and spore types in pollen trap samples from the Ulu Gombak Virgin Jungle Reserve, Selangor, W. Malaysia. (After Flenley, 1973)

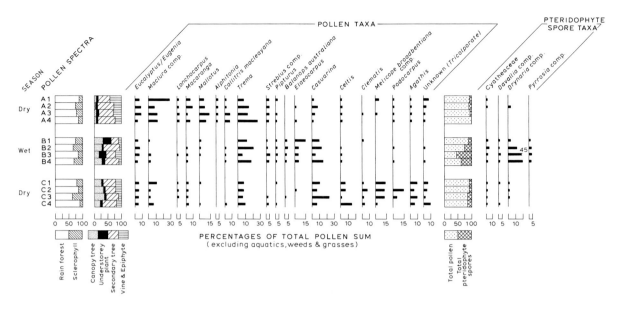

Figure 5.13 Pollen diagram of best represented pollen and pteridophyte spore taxa and ratios between selected taxon groups from trap samples, Atherton Tableland, Queensland, Australia (A1, A2 etc. are unroofed traps; A3, A4 etc. roofed traps). (After Kershaw and Hyland, 1975)

Modern pollen studies have also been carried out in the Queensland rain forest (Kershaw, 1973b; Kershaw and Hyland, 1975). Tauber traps were placed floating on Lake Euramoo, at least 80 m from the forest edge. Some traps were roofed to exclude rain-borne pollen, others were exposed to rain. Despite difficulties caused by the addition of extraneous matter, chiefly by trap-perching cormorants, Kershaw and Hyland (1975) examined spectra from the traps for two dry seasons and the intervening wet season. The results (*Figure 5.13*) suggested that distinct wet and dry season spectra existed and pollen deposition was greatest at the end of the dry season. Absolute deposition was only 90–250 grains per cm^2 per annum, very much lower than within the Malaysian rain forest. Of course, this may be due as much to poor transport over the 80+ m from the forest, as to low production by the forest. The pollen trapped appeared to consist largely of that from canopy taxa, suggesting that little trunk-space pollen was carried out on to the lake. Roofed and unroofed traps had similar amounts and types of pollen in them suggesting the rain component is small.

It is too early to draw conclusions about lowland forest pollen rain, but the following suggestions may be tentatively offered:

1. The chief unusual feature of the pollen rain is that it is very diverse, although not impossibly so from the point of view of the pollen counter. Identifications frequently have to be only to the family or genus, which will make ecological interpretation of pollen diagrams very difficult.
2. The magnitude of the total pollen rain is lower than in temperate regions, though perhaps not markedly lower.
3. Lateral transport of pollen seems to be relatively poor.

4. A distinct canopy component appears to exist; the trunk-space component may exist but probably does not pass outside the forest significantly; the rain component may be small.

Bearing in mind all these findings and other unpublished data (Powell, 1970; Hope, 1973; Morley, 1976; Maloney, in preparation; J. Muller, personal communication) we may make preliminary generalisations about some of the major pollen types which occur in the pollen diagrams from Indo-Malesia, and these are given in Appendix 3.

5.5   THE FOSSIL POLLEN EVIDENCE

5.5.1   NEW GUINEA

The best evidence so far comes from Sirunki swamp at 2500 m (*Figure 5.14*). This remarkable site has given a continuous record to beyond the range of carbon dating (D. Walker, personal communication) and pollen analysis (*Figure 5.15*) back to c. 33 000 B.P. has been carried out (*Figure 5.16*). The diagram is unusual in that results have been expressed as grains per cm^2 per annum, assumed to be the pollen deposition rate (PDR). The inferred ages (I.A.) given are based on a fairly large number of radiocarbon dates and other information, which distinguishes them from the usual uncorrected carbon dates given by most authors and designated B.P.

Interpretation is based not only on detailed consideration of the taxa represented but also on Hope's (1973) results from pollen trapping on Mt Wilhelm, which showed that in unforested catchments the PDR was often less than 2000, whereas in forested catchments a PDR consistently over 2000 and often over 4000 could be expected. Since the diagram is more

Figure 5.14 An oblique aerial view of Lake Iviva (Ipea) and Sirunki Swamp, New Guinea, c2500 m. The ground surrounding the swamp bears Miscanthus grassland; the ridges bear Nothofagus forest. The pollen diagram Figure 5.16, comes from a core taken beyond the road which crosses the swamp. The diagram covers the last 30 000 years and shows multiple changes in vegetation

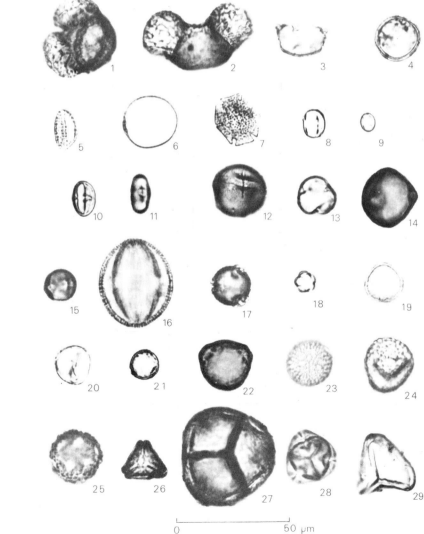

Figure 5.15 Fossil pollen and spores from the Quaternary of New Guinea. All the specimens shown are from Lakes Inim and Birip

1. Dacrycarpus
2. Podocarpus
3. Phyllocladus
4. Ascarina
5. Astelia papuana
6. Gramineae
7. Nothofagus (brassii-type)
8. Saurauia
9. Elaeocarpus
10. Lithocarpus/Castanopsis
11. Oreomyrrhis
12. Coprosma
13. Rapanea
14. Dodonaea
15. Macaranga
16. Gentiana
17. Potentilla
18. Quintinia
19. Trema
20. Engelhardia
21. Acalypha
22. Casuarina
23. Drapetes
24. Plantago (aundensis-type)
25. Ranunculus
26. Myrtaceae
27. Ericaceae
28. Epacridaceae
29. Cyathea
(after Flenley, 1967)

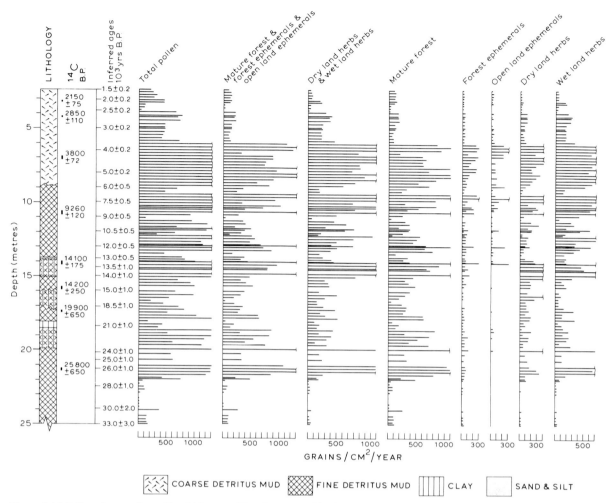

Figure 5.16 Pollen diagram from Sirunki Swamp, New Guinea Highlands. Values are pollen deposition rates in grains per cm² per annum. Note changes of scale. Only selected taxa are shown. The diagram suggests almost continuous vegetational change throughout the last 33 000 years but indicates particularly depression of the forest limit between 24 000 and 14 000 B.P. and forest clearance since 4500 B.P. (After Walker and Flenley, in press)

an exhibition of continuous change than of stable phases it has not been formally zoned, but it is never-theless convenient to discuss the interpretation in time bands, as follows:

33 000–27 500 I.A.

PDR is so low as to indicate certainly unforested, indeed hardly vegetated, conditions.

27 500–25 500 I.A.

Forest is indicated by the high PDR values, and the abundance of pollen of Nothofagus, Castanopsis, Phyllocladus etc. These are taxa of lower mountain forest today. Rapanea pollen and Cyathea spores, also abundant, suggest upper mountain forest nearby, whilst the presence of Astelia pollen argues for sub-alpine swamps.

25 500–24 000 I.A.

The PDR values decline markedly suggesting demise of the forest; grasses and Astelia dominate the locally produced pollen.

24 000–14 000 I.A.

The fluctuating, but usually low, values for PDR suggest generally unforested conditions, and a veg-etation of wetland and dryland herbs. The values for wetland herbs commonly just exceed those for dryland herbs before about 18 500 I.A., but the reverse obtains after that date.

14 000–13 250 I.A.

This appears to have been a time of dramatic change. Lower mountain forest clearly invaded the catchment, with upper mountain forest on the higher slopes (poss-ibly above 2800 m). Nothofagus appears to have been common, with increasing participation from Pod-ocarpus and later Castanopsis and Phyllocladus. The presence of forest ephemerals (pioneers) such as Macaranga and Trema at this time is interesting, al-though not all New Guinea Macaranga spp. are pioneers (Flenley, 1969).

13 250–9000 I.A.

The values for PDR fall off again in this phase, suggesting forest had disappeared from the catchment by 12 500 I.A., being probably replaced by *Cyathea* 'woodland', sedge swamps and grassland. There is stratigraphic evidence for a rise in water level culminating about 13 000 I.A. *Astelia* was apparently present in the catchment until 10 000 I.A., but probably not thereafter.

9000–8000 I.A.

This was essentially a time of forest development. Early representatives were *Nothofagus, Ascarina, Castanopsis, Phyllocladus* and *Podocarpus*. High values for *Macaranga, Saurauia, Trema* and *Dodonaea* imply that these probably behaved as pioneer species, and that it took 1000 years for closed forest to be established. At this time a rapid overgrowth of the Sirunki Lake with swamp vegetation probably began, to be completed about 6000 I.A.

8000–5000 I.A.

This was apparently a time of forest dominance, and for the last 1500 years of it the forest appears relatively stable, dominated by *Nothofagus*.

5000–4000 I.A.

This millenium apparently saw a decline in *Nothofagus* forest, particularly after 4300 I.A. There are large increases in the representation of *Trema* and *Acalypha*, suggestive of open conditions.

4000–1500 I.A.

During this part of the diagram there are pronounced reductions in PDR which possibly relate to the overgrowth of the sample site by reedswamp, and the results are therefore difficult to interpret. They contain, however, nothing to suggest otherwise than the continuation of the mixture of forest and open vegetation which was already established by 4000 I.A., and which occurs today in the area.

These changes have been generalised as follows (Walker and Flenley, in press):

'In summary, the nature of the vegetation in the catchment before 27 500 I.A. is enigmatic, although the weight of evidence favours alpine barrens. From then until 9000 I.A. subalpine conditions dominated and trees and shrubs were sparse except for the periods 27 500–25 500 I.A. and 14 000–13 250 I.A. which witnessed excursions of lower mountain forest taxa into the catchment. The final afforestation began about 9000 I.A. and attained the dominance of a *Nothofagus*-rich forest by 6500 I.A. which was maintained for 1500 years. However, at 5000 I.A. fluctuations in forest composition began which were associated with periodic changes in the proportion of forested to unforested land and which certainly directly affected the catchment from 5400 I.A.'

That these changes, or at least the later stages of them, were no local anomaly, is attested to by several other pollen diagrams. A percentage diagram from Lake Inim (*Figure 5.17*) at 2550 m (Flenley, 1967, 1972; Walker and Flenley, in press) is summarised in *Figure 5.18*. This site is very close to Sirunki, and a similar history might be expected. Unfortunately the dating of the deposits is not satisfactory, but the diagram clearly suggests an open subalpine or alpine vegetation (*sensu* Wade and McVean, 1969) in the earlier time represented, and forest more recently. The boundary between these two may be placed at approximately 8000–12 000 B.P. which covers the date of forestation at Sirunki. The diminution of forest at 5000–4000 I.A., so clear in the Sirunki diagram, is apparently a diachronous event, for at Inim it does not begin until an estimated c. 1600 B.P. This and other

Figure 5.17 Lake Inim, New Guinea, 2550 m. The swamp immediately around the lake is floating. Most of the swamp is dominated by the sedge Machaerina rubiginosa. In the foreground on dry land are the grass Miscanthus floridulus and gardens of sweet potato Ipomoea batatas, with a relict patch of Nothofagus forest at the right. The site has yielded a pollen diagram, Figure 5.18, showing that the area was formerly forest covered and earlier still, before 12 000 B.P., bore tropicalpine vegetation. (Photograph by J. R. Flenley)

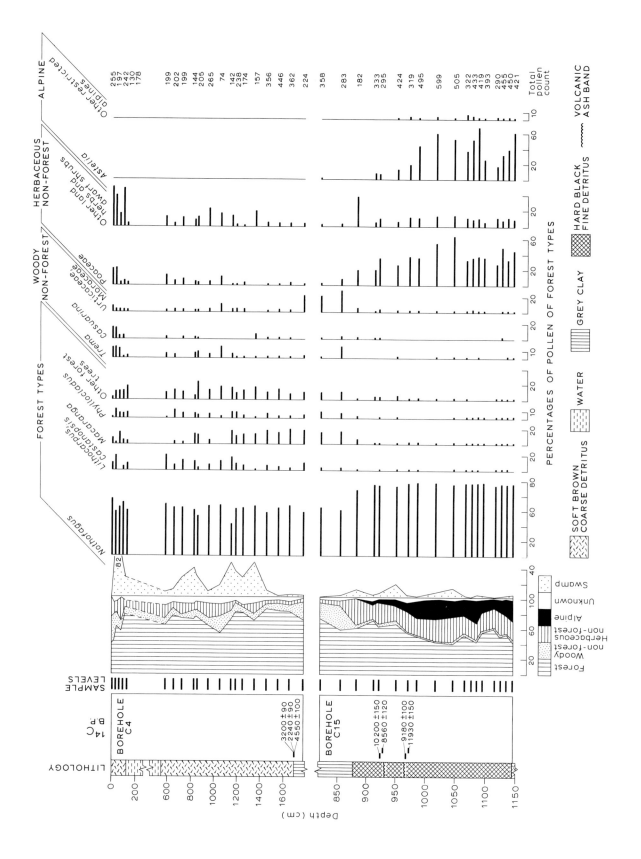

Figure 5.18 Pollen diagrams from Lake Inim, New Guinea Highlands, boreholes C4 and C15, plotted on the same scales. The results are expressed as percentages of pollen of forest types, except in the summary diagram where total dry land pollen and spores is the pollen sum. Only selected taxa are shown. There is clear evidence of former depression of the forest limit. (After Flenley, 1967; 1972, and Walker and Flenley, in press)

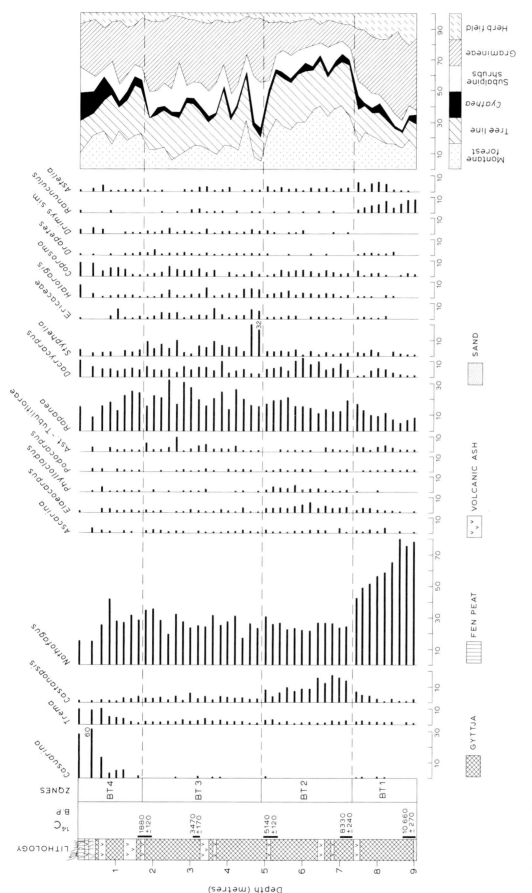

Figure 5.19 Pollen diagram from Brass Tarn, Mt Wilhelm, New Guinea (altitude 3910 m). Only selected taxa are shown. Values are percentages of total pollen for all woody plants excluding Casuarina. The maximum values for montane forest in zone BT2 can be clearly seen in the summary diagram at the right hand side. (After Hope, 1976a)

evidence suggests that the recent deforestation is a man-made event (*see* Chapter 7).

The suggested presence of 'alpine barrens' at Sirunki prior to 27 500 I.A. is difficult to explain in terms other than those of a much more severe climate than at present. Such an occurrence might be expected to be reflected at lower altitude. It is therefore exceedingly interesting to note than in the pollen diagram from Draepi Swamp (Powell, 1970) at only 1885 m shown in *Figure 7.8*, there is a phase (Zone G) in part of which Podocarpaceae assumes importance and even *Astelia* is present, suggesting the proximity of some subalpine vegetation; this phase is of inexact date but prior to 30 000 B.P. Similar evidence is available from sites at about 1500 m, further south in New Guinea (Williams *et al.*, 1972; Powell in press b).

Confirmatory evidence from further afield comes from the elegant suite of sites on Mt Wilhelm (Hope, 1976a). The major aim of this work was to trace the deglaciation of the mountain and the movements of the tree limit during the last 20 000 years. The results, summarised in *Figure 5.23*, show rather clearly that the tree line was lower than 2740 m before c. 10 800 B.P. and then rose gradually (with considerable individualism shown by taxa) reaching beyond 3910 m about 8500 B.P. Later, however, a decline appears to have set in and forest retreated, from about 5000 B.P., presumably to about the present tree line around 3750–3900 m. This last change is not well shown elsewhere, and is therefore of considerable interest. The pollen diagram which is taken to indicate it is that from Brass Tarn at 3910 m, shown in *Figure 5.19*. It is not clear whether this change should be attributed to climatic change, human impact, or some other factor. The usual pollen indicators of disturbance (*Trema, Casuarina* etc.) are only abundant after c. 2000 B.P., and are in any case often carried up from lower altitudes. Firing by hunting parties cannot, however, be ruled out at 5000 B.P., and the cause of this change remains uncertain.

Tree-line movements have also been investigated in Irian Jaya (W. New Guinea) where Hope and Peterson (1976) found pollen evidence at 3630 m that subalpine grasslands with scattered shrubs had given way to subalpine forests by 10 750 B.P. The forests subsequently declined, since perhaps 5000 B.P., leaving the present mixture of forest, shrubland and grassland (Hope, 1976b).

The consistency of these New Guinea findings is remarkable. The types of vegetational change noted are not, with the exception of some in the last 5–10 000 years, the sort likely to be induced by man. *In toto* the results argue for major changes which can only be explained in terms of climatic variation of the most marked kind.

5.5.2   JAVA AND SUMATRA

Some evidence relating to Quaternary vegetation is available from Sumatra and Java. The Javanese evidence is that of macrofossils from the Middle Pleistocene deposits at Trinil (Schuster, 1911 a,b). Modern palynological studies of deposits from the same sequence are being undertaken by van Zeist and co-workers at Groningen, and the results are awaited with great interest. The particular significance of these deposits is that they are those in which 'Java man' (*Homo erectus*) was found, and the plant fossils therefore tell us something of the environment of early man as well as providing data for the construction of a vegetational history. It is important to realise that the actual stratigraphy at Trinil is complex, and that the plant fossils do not come from the same bed as the fossil bones (which include many other animals as well as *Homo erectus*), but from slightly above it. The age of these deposits is Early Pleistocene, possibly back to 1·7 M B.P. (Lestrel, 1976).

The determinations by Schuster (1911 a, b) are chiefly based on fossil leaves and are unsupported by cuticular studies, so they must be regarded as rather unreliable. Nevertheless, they are unlikely to be all wrong, and the general relationships of the assemblage are quite clear: they are mainly lowland rain forest species. Even if the determinations are disregarded, a high proportion of the leaves figured by Schuster have entire margins, and several show drip-tips; these facts support the rain forest origin of the assemblage. This is an interesting conclusion for the present vegetation of the area, in so far as it has not been destroyed by man, is teak forest (van Steenis and Schippers—Lammertse, 1965). As the deposit is a fluvial one, however, it is not clear how far the material might have travelled, so that a conclusion that vegetational change has occurred cannot be strongly maintained. Still less is it possible to conclude that this was the environment of Java man, since the flora is not from the same horizon as the hominid fossils.

The evidence from Sumatra relates to quite a different time period, namely the last 11 000 years. Two pollen diagrams have been prepared from the Kerinci Valley in Central Sumatra (Morley 1976, in press; Flenley, unpublished observations). The aim was to obtain evidence of the history of the lowland rain forest, and sites were, accordingly, chosen at 950 m and 1050 m, covering the boundary at about 1000 m between lowland and sub-montane* forest in the area. The diagram from Danau Padang at 950 m (*Figure 5.20*), which is the first diagram from lowland Malesia to cover virtually the complete Holocene, indicates two main phases. The later phase (c.8000 B.P. to present) is dominated by pollen of wide altitudinal range, but with a fair sprinkling of taxa confined to sub-montane forest. Reference to the modern pollen rain from the area (*Figure 5.10*) will show that such spectra are typical of the vegetation around 950–1000 m i.e. the actual altitude of the site. This part of the diagram therefore suggests a forest composition of, broadly speaking, present-day type. The earlier phase (c.9800–8000 B.P.) is again dominated by pollen of wide altitudinal range, but this time with a

---

* The forest types defined by Morley (1976) are not identical to those defined for Malesia generally (e.g. by Whitmore, 1975).

DANAU PADANG. Lat. 2° 15'S. Long. 101° 31'E.

Pollen sum total primary arboreal pollen

Selected pollen types only

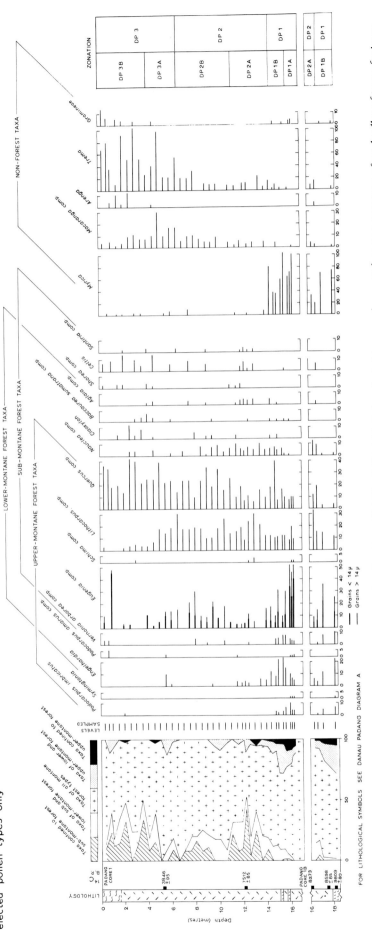
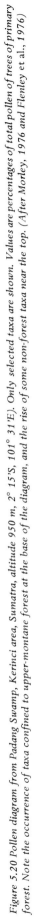

Figure 5.20 Pollen diagram from Padang Swamp, Kerinci area, Sumatra, altitude 950 m, 2° 15'S, 101° 31'E). Only selected taxa are shown. Values are percentages of total pollen of trees of primary forest. Note the occurrence of taxa confined to upper-montane forest at the base of the diagram, and the rise of some non-forest taxa near the top. (After Morley, 1976 and Flenley et al., 1976)

Figure 5.21 Quincan Crater, North-east Queensland, Australia, c.790 m. The crater walls still bear relict rain forect trees. The pollen diagram shows that rain forest migrated into the area 6000 years ago, replacing an earlier vegetation of Eucalyptus. *(Photograph by A. P. Kershaw)*

significant proportion of taxa from Morley's upper montane forest e.g. *Podocarpus* (*Dacrycarpus*) *imbricatus* and *Symingtonia*, or present in both upper and lower montane forests, e.g. *Engelhardia* and *Podocarpus amarus*. Comparison with modern pollen rain suggests that this kind of spectrum might be obtained today only above 1700 m, and even allowing for possible transport of pollen from low hills around the site suggests an altitudinal shift in the vegetation zones of at least 500 m in this area. This is a fairly high value for such a low altitude. The pollen diagram from Sikijang Swamp at 1050 m (Flenley, unpublished observations) is broadly similar to the Danau Padang diagram (though extending back a little further, to 11 000 B.P.) and provides evidence that these were regional changes rather than local anomalies.

It is theoretically possible that the two tectonic sites from which the pollen diagrams were taken had existed long previously but only began to accumulate sediment when water accumulated in them. This might imply a gradually moistening climate around 11 000–10 000 B.P. However, a third core from a crater further north in Sumatra, at present under analysis (Maloney, in preparation) has a continuous record back to 18 000 B.P. with no pollen evidence suggesting dry conditions.

5.5.3   SOUTH INDIA

There are a few pollen diagrams available from southern India (Gupta, 1971; Vishnu-Mittre and Gupta, 1971)

and radiocarbon dates back to beyond 25 000 B.P. promise considerable interest. In the absence of modern pollen rain studies in the area, however, it is very difficult to interpret these diagrams. It seems clear that they show significant changes in the former vegetation, but detailed conclusions must await further work.

5.5.4   QUEENSLAND, AUSTRALIA

Despite its situation at 17–18° S, well outside the main equatorial area, the Atherton Tableland of Queensland is mentioned here, because it has given rather clear evidence of Quaternary vegetational change. Cores from four sites near the present boundary between rain forest and sclerophyll woodland (*Figure 5.3*) have been analysed. Three of these, Lake Euramoo (Kershaw, 1970a), Quincan Crater (*Figure 5.21*) (Kershaw, 1971) and Bromfield Swamp (Kershaw, 1975a) each cover at least the last 8000 years. Lynch's Crater (Kershaw, 1975b, 1976) gives evidence of vegetational changes from about 60 000 B.P. to less than 10 000 B.P. Simplified pollen diagrams from Lake Euramoo and Lynch's Crater (Kershaw, 1975b) between them cover the whole time span of record available, and contain rather clear changes in pollen spectra (*Figure 5.22*). Bearing in mind the modern pollen rain record from the area (*Figure 5.13*), zones L5–L3 may be interpreted as reflecting the existence of rain forest of some sort close to Lynch's Crater, although sclerophyll woodland may have been not far

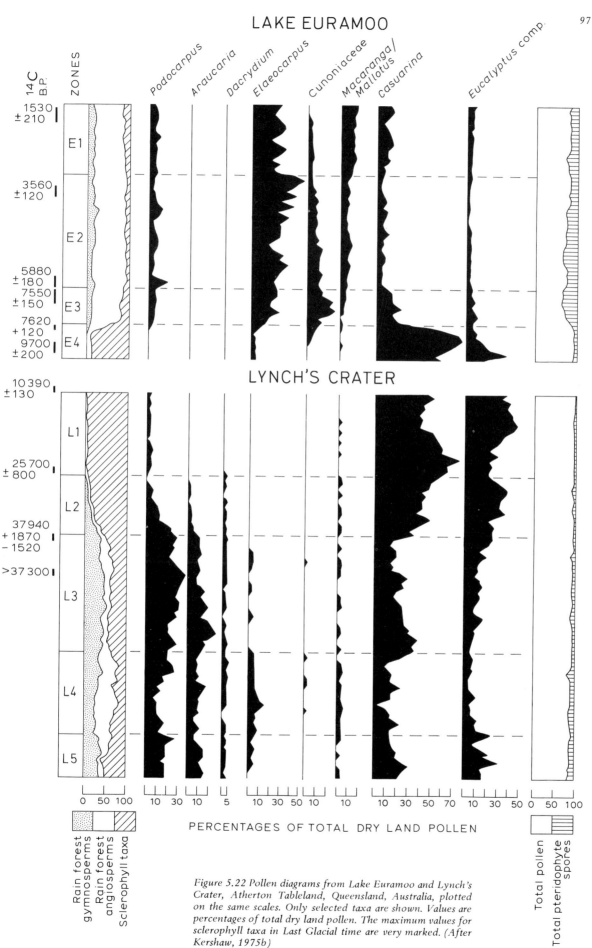

Figure 5.22 Pollen diagrams from Lake Euramoo and Lynch's Crater, Atherton Tableland, Queensland, Australia, plotted on the same scales. Only selected taxa are shown. Values are percentages of total dry land pollen. The maximum values for sclerophyll taxa in Last Glacial time are very marked. (After Kershaw, 1975b)

away. Apparently this rain forest was gymnosperm-dominated; the approximate name according to Webb's (1968) classification is vine forest plus *Araucaria*. Zone *L2* (c.38 000 to c.26 000 B.P.) can be regarded as a transitional phase when the rain forest declined and eventually totally disappeared from the area, being replaced by sclerophyll woodland. The sclerophyll vegetation persisted until the end of Zone *E4* (c.7500 B.P.) when rain forest returned abruptly to the area, although this time without appreciable gymnosperm representation. This vegetation, in general, persisted until the present although there was again some variation in the precise type of rain forest. It is an important point that the return of rain forest to the area is documented also at Bromfield Swamp (where it occurred between 9500 and 8400 B.P.) and at Quincan Crater (where it is dated to 7000–6000 B.P.). These slightly differing dates strongly support the idea that an actual migration of rain forest was occurring over the period covered through a distance of at least 20 km.

The precise causes of these vegetational changes are not certain. The decline of the early gymnosperm-dominated rain forest since 38 000 B.P. was, it must be suspected, due to a reduction in available moisture. Kershaw (1975b) points out, however, that a feasible alternative explanation would be the lighting of fires by aboriginal man, who is known to have been in Australia from at least 32 000 B.P. (Barbetti and Allen, 1972), and might well have been there from 38 000 B.P. when the reduction in rain forest began. Although this may well have been contributory, that it was the sole cause seems to be a little unlikely, in view of the re-appearance of rain forest between 9500 and 6000 B.P. at three different sites. It is difficult to see why aboriginal fires should have ceased to be effective at this time, and the likely cause of this later change is a return to moister conditions. One cannot fail to notice the general correlation of the presumed dry period (indicated by sclerophyll woodland) with the period of savannah vegetation in South America and Africa, and with the time of maximum glaciation in temperate regions. This provides further evidence against the old 'pluvial' theory. Kershaw (1975b) points out that the record of organic deposition at three of the four sites only begins in the period 11 000–7000 B.P., and suggests that prior to this the volcanic craters in which these sites lie were in existence but without permanent water. He therefore argues for a progressive increase in effective precipitation over the period 11 000–7000 B.P. One anomaly remains: why was the 'later' rain forest lacking in the gymnosperms which dominated the 'earlier' rain forest? As regards *Araucaria cunninghamii* at least, one possible explanation is its slow migration rate. Havel (1971) showed that its seeds can usually travel not more than 60 m from the parent tree, so the species would obviously have required thousands of years to migrate out from its few Queensland refugia and re-colonise all the areas potentially available. Refutation or confirmation of many of these ideas may come from the extension of

this record back into the time of the last interglacial of temperate areas. Lynch's Crater with organic sediments exceeding 40 m, and only 20 m so far analysed, indeed promises a record of this length, and academic tongues hang out in eagerness for the future results (*see* Kershaw, in press).

It should be pointed out that if precipitation were less in North Queensland during the period 38 000–7500 B.P., this was not necessarily related only to reduced evaporation from the oceans resulting from lower global temperatures. The explanation is more simple if eustatic changes in sea level are also invoked. Assuming the Sahul shelf to the north-west of Australia was exposed at this time, and there is evidence that it was (van Andel and Veevers, 1967; van Andel *et al.*, 1967), the north-west winds which at present provide moisture for several months of the year would have lost much of their moisture while passing over the shelf land areas, and even the south-east trades which are the main bearers of rain would have had to cross an additional 60–80 km of plain (Nix and Kalma, 1972; Webster and Streten, 1972). Combined with the presumed reduced evaporation from the oceans at the time, this would have been likely to cause a marked reduction in precipitation.

5.6   CONCLUSIONS

5.6.1   VEGETATION

The conclusion now seems inescapable that the montane vegetation of New Guinea was profoundly different during the Late Pleistocene from what it is now, and that this difference was in general, although not in detail, a depression of present vegetational zones at roughly the same time as those in Africa and South America were depressed. The evidence for this is summarised in *Figure 5.23*. The amount of maximum depression of the forest limit – c.1000 m to c.1500 m – was also roughly comparable with that at similar times in other equatorial areas. The evidence from lower altitudes in Sumatra, although not yet fully conclusive, does not conflict with that from New Guinea. In Sumatra an altitudinal lowering about 9000 B.P. of c.500 m was postulated for the lower limit of the montane forest. This may be compared with the evidence from New Guinea in the summary diagram for that area (*Figure 5.23*).

It has been pointed out by Walker (personal communication), however, that it might be more accurate to speak not of the depression and re-elevation of the upper mountain forests in New Guinea, but rather of their destruction and subsequent re-formation. The chief evidence for this idea is in the pollen diagrams. At Sirunki, Inim and Komanimambuno (the lowest site on Mt Wilhelm) there is little evidence for the existence of a substantial sub-alpine forest in the Late Pleistocene, and even the mixed mountain forest is scarcely indicated. If these forests had been present, even well below the pollen sites, their pollen would be expected to have been preserved. It is clear from

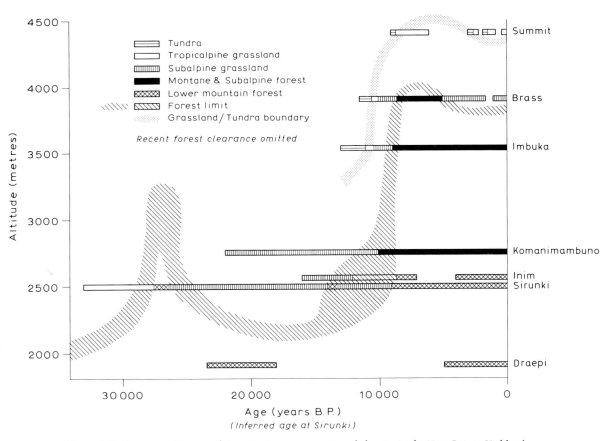

Figure 5.23 Summary diagram of Late Quaternary vegetational changes in the New Guinea Highlands. (Partly after Hope, 1976a)

the Holocene of the Mt Wilhelm sites that both these forest types can show clearly in pollen diagrams. Their absence suggests that perhaps they did not have distinct refugia in the Pleistocene, but that their separate specific elements survived as rare components of the upper fringes of the mountain *Nothofagus* forests, and amalgamated during the Holocene to form the present distinct communities.

This suggests a large amount of individualistic behaviour by the forest species, which was apparently also the case in the lower mountain (beech and oak) forests. To some extent this is suggested by the pollen diagram from Sirunki (Walker and Flenley, in press), but it is also suggested by the variability of these forests today. Each mountain has its own peculiarities of flora and dominance, and these do not correlate well with edaphic or other present ecological factors (Flenley, 1969). This could be explained if, when large montane areas were, about 10 000 years ago, rather suddenly available once again to mountain forest species, the latter migrated in not as a single community, but as individual species, so that first arrivals gained an advantage, just as weed species may do in an area of cleared land. Due to the relatively small number of tree generations in the last 10 000 years, and the isolation of the forests on separate peaks, the reverberations of these historical accidents have not yet been damped down (Flenley, 1969). The extent to which this idea is valid, and whether it can be extended to other communities, is still under investigation.

The evidence from Queensland suggests that Late Pleistocene vegetation there was also strikingly different from that in the last 6000 years. If the change from sclerophyll to rain forest had occurred at a single site, it could have been regarded as a local anomaly, but to find the same event at four sites strongly suggests a regional change; the fact that the event was not synchronous everywhere can be explained as a migration in response to a gradually changing climate. Why, then, should the vegetational change, which is presumably a response to an increase in moisture, have occurred in Queensland when there was no corresponding hydrologically induced vegetational change in New Guinea or Sumatra? Firstly, the difference in latitude implies that no parallel change should necessarily be expected. Secondly, the marginal situation of the Atherton Tableland in Queensland for rain forest growth, has probably made its vegetation exceptionally sensitive to climatic change, whereas areas with climates closer to optimum rain forest conditions could survive considerable relative dessication without significant change in their vegetation.

5.6.2 CLIMATE

The vegetation evidence from New Guinea is concordant with a reduction in temperature during the period 18 000–16 000 B.P. of between 7 and 11 °C

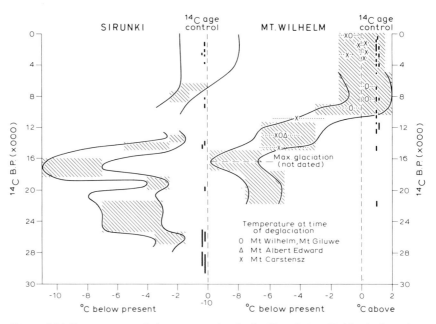

Figure 5.24 Temperature variations at two sites in the New Guinea Highlands throughout the last 28 000 years, as suggested by pollen analysis and glacial evidence. The boxes and double curves indicate areas within which the true values almost certainly lie. (After Bowler et al., 1976)

(*Figure 5.24*). Between 18 000 and 28 000 B.P., temperatures were possibly below present values, but by only 2–8°C. After 16 000 B.P. there is evidence of a transition, possibly with some oscillations, reaching present day temperatures by 9000–6000 B.P. The Sumatran evidence suggests a possible temperature of 2–4°C below present values in the period 11 000–8000 B.P., which does not conflict with the New Guinea evidence. The Queensland evidence, although not demanding interpretation in terms of temperature change, does not necessarily exclude cooler temperatures in the period c.11 000–7000 B.P., and possibly back to 38 000 B.P.

In neither Sumatra nor New Guinea is there yet any Quaternary vegetational evidence which can *only* be explained in terms of a formerly *drier* climate. This is not to say that precipitation may not have been reduced. The present climates of the areas studied are abundantly moist, and it is likely that precipitation could be reduced, or for that matter increased, by as much as 30%, without radically altering the vegetation, unless there were also a pronounced change in seasonality. Such a reduction might have occurred in New Guinea due to the emergence of the Sahul shelf at times of eustatically lowered sea level (Nix and Kalma, 1972) and in Sumatra due to the similar emergence of the Sunda shelf (van Steenis, 1935; Ashton, 1972). In Queensland the indication of drier climate during the period 26 000 to c.7000 B.P. is rather clear.

5.6.3   BIOGEOGRAPHICAL PROBLEMS

The preliminary evidence of Quaternary vegetation in Indo–Malesia goes some way towards explaining the occurrence of mountain species on separate peaks. When the forest limit, for example, was depressed to only 2000 m, the area open to tropicalpine vegetation was tremendously increased and many now separate tropicalpine areas were joined. Clearly the great increase in potential habitat would have provided almost continuous corridors for the migration of those species. Such findings do not, of course, account for really major problems of disjunction, but the expanded area available to many stenotherm taxa in the Quaternary, could have provided a greater source of propagules for long distance dispersal.

On the problem of possible connections between the vegetation of Thailand and East Java (and the Lesser Sunda Isles), vegetational history has as yet no direct evidence. The changes found in Queensland do, however, suggest that it is worthwhile to look in the lowlands near the equator for evidence of former expansion of savannah vegetation, and this must surely be one of the major problems currently worth investigation in the region.

The problem of Wallace's line is not yet solved, and it may well be that vegetational history will not provide the major contribution to explaining why the boundary is of more significance to some taxonomic groups than to others. Perhaps the answer will be in studies of dispersal power and modes of dispersal of individual taxa, along with ecological studies of the establishment of propagules.

6
Seral Changes in Equatorial Vegetation

6.1 INTRODUCTION

The successional or seral changes which occur, or are believed to occur, in equatorial vegetation may be conveniently grouped under two main headings. The first group are xeroseres, i.e. those which occur on dry land; the psammosere, the succession on coastal dunes, may be regarded as a particular type of xerosere. The second group are the hydroseres, occurring in fresh water; the mangrove swamps of equatorial coasts will, for convenience, be considered along with these.

The classical views on these successions have been outlined in Chapter 1, and are dealt with in more detail in well-known works such as that of Richards (1964), so that there is little point in repeating them here. One of the difficulties in studying seral changes, as clearly pointed out by earlier workers, is that it is difficult for the observer to reconstruct accurately the course of changes which may be occurring very slowly. Seral vegetation is usually zoned and the common problem is 'Does the zonation represent a succession?' Although researchers have always been aware of this problem, earlier studies, in temperate regions particularly, often did lead to the conclusion that zoned vegetation was indeed the expression at one phase in time, of an underlying succession. Recently, however, this has frequently been questioned. For example, Walker (1970b) showed the hydrosere in British fresh waters did not necessarily follow the 'classical' pattern. Stratigraphic evidence, using both macrofossils and pollen, showed that there was no stage when a hydrosere was committed to a particular future sequence of changes; rather, each successive change depended on a variety of factors from which random chance could by no means be excluded. In the tropics, the view that the mangrove swamp is migrating seawards as it stimulates accretion and beach development (i.e. that these processes are autogenic) has been seriously questioned by Scholl (1968) who states that 'Mangroves do not significantly increase the rate of seaward or lateral growth of the coast'. This view is shared by Thom (1967) working on Mexican mangroves, and by Hopley (1970) who studied similar vegetation on the coast of Queensland, Australia.

These difficulties mean that in order to establish the succession through time for a particular point or area in space it is essential to have either historical or stratigraphic records. Macrofossils are usually preferable to pollen as stratigraphic evidence in this case, because the area from which the pollen is derived may well cover several vegetational zones.

6.2 XEROSERES

Evidence about tropical xeroseres is rather scanty, chiefly because the dry-land conditions, which by definition prevail, are not conducive to the preservation of a good stratigraphic record. Usually we have to rely on a nearby hydroseral record which accidentally incorporates xeroseral material. For example, there are a few cases where the development of vegetation on a newly deglaciated land surface has been documented in the deposits of a tropical mountain lake. In the

Ruwenzori, East Africa, the pollen diagram from Mahoma Lake (Livingstone, 1967), *Figure 3.14*, shows an early abundance of Ericaceae, followed by Gramineae and Cyperaceae.

These can be interpreted as seral vegetation, or as a vegetation controlled at this level by severity of climate. Livingstone prefers the latter explanation, but regards the arrival of forest shortly afterwards as due to maturing of soils rather than climate change.

In the Andes, Gonzalez *et al.* (1966) record the vegetational history of the Valle de Lagunillas as deglaciation proceeds. The pollen diagram from Core VL–XI (3990 m) has a basal sample with abundant fern spores, followed by high values for Gramineae, and a similar sequence occurs in the diagram from VL–X (4220 m), but the data are too exiguous for firm conclusion. In New Guinea, on Mt Wilhelm, lithoseral stages are reported (Hope, 1976). The basal two samples of the pollen diagram from Brass Tarn, for example, (*Figure 5.19*) are rich in pollen of *Ranunculus*, but have only moderate amounts of Gramineae (Poaceae), and probably represent a 'tundra' vegetation. The subsequent samples are much richer in pollen of Gramineae, while *Ranunculus* pollen is less abundant, and some kind of 'alpine' grassland is indicated. The change from tundra to grassland may have been seral, as may the subsequent arrival of forest.

There is much better evidence, both as recorded history and from palynology, for another type of xerosere, the vulcanosere. It will come as no surprise that this evidence is from the Indo–Malesian region which is so well endowed with volcanoes. From the point of view of the biogeographer, volcanoes are roughly divisible into those that produce lava when they erupt, those that produce scoria (solid material of gravel or larger size) and those that produce fine volcanic ash. Actually the last two of these are normally produced simultaneously but the scoria drop onto the sides of the volcanic cone, while the ash may be carried very much further through the air. Additional differences are that lavas often tend to be rich in basic minerals, which are therefore liable to yield on weathering an alkaline or neutral soil, whereas scoria and ash, particularly the latter, consist commonly of quartz and other acidic minerals likely to give rise to acid soils.

The classic case of Krakatau is so well known as to need little description. Unfortunately there is some doubt as to whether the eruption of 1883 did really destroy all the original vegetation (Backer, 1929), although as Richards (1964) concludes, there is a strong probability that it did. Certainly there can have been very few species left, and the steady arrival of new species since then has been well documented by Docters van Leeuwen (1936), as summarised in *Figure 6.1*. It is interesting to note that the rate of increase in number of species shows no signs of falling off with time, as might be expected on theoretical grounds when extinctions begin to occur due to competition (MacArthur and Wilson, 1967). These authors offer two possible explanations:

'either the pool of plant species on Java and Sumatra is so enormous in comparison with the number of species settled on the Krakatau islands by 1934 that a depletion effect is not visible, or else the extinction curve actually *declined* with species buildup. The flora of the greater Sunda islands is certainly enormous enough to mask a depletion effect, although the pool of effective colonisers among the plant species may not be. The second explanation seems equally reasonable. When plants as a whole are colonising a barren area, the extinction curve should decline at first, due to succession. Later plant communities are dependent on earlier, pioneer communities for their successful establishment. Yet when they do become established, they do not wholly extirpate the pioneer communities, at least not if the sample area is topographically varied enough; and a large part of the total successional diversity is thereby preserved. Furthermore, trees do not grow tall for many years, and during this interval no deeply shaded forest can be found.'

The arrangement of all these immigrating species into successional communities has been summarised, with some assumptions, by Richards (1964), as shown in *Figure 6.2*.

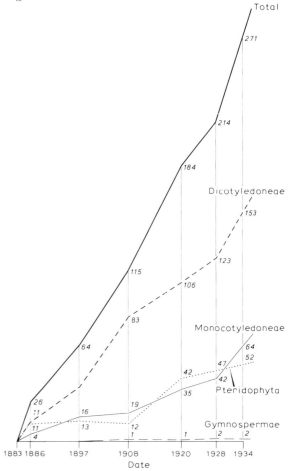

Figure 6.1 Buildup of numbers of species of three higher plant groups on the three islands of the Krakatau group since the eruption of 1883. (After MacArthur and Wilson, 1967)

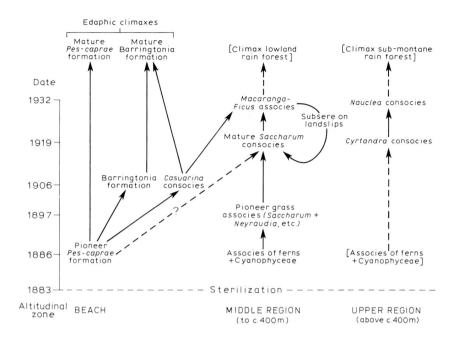

Figure 6.2 Diagram of successions on Krakatau since the eruption of 1883. (After Richards, 1964)

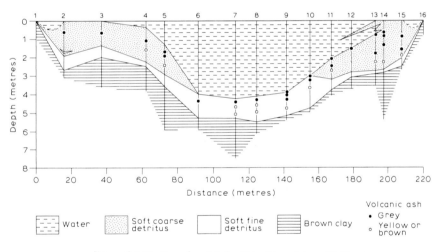

Figure 6.3 Stratigraphy at Lake Birip, New Guinea Highlands.
(After Flenley, 1967; Walker and Flenley, in press)

Table 6.1. REVEGETATION OF VOLCANIC BLAST AREAS IN NEW GUINEA (AFTER TAYLOR, 1957 AND WHITE, 1975)

Volcano	Blast Area (km²)	Year of Eruption	Years Since Eruption	Secondary Species	Climax Species	Total	Tree Species In Richest Sample (2 ha)
LAMINGTON	200	1951	6	26	0	26	10
WAIOWA	40	1944	13	17	8	25	15
VICTORY (Lowland zone only)	300	1870?	87	54	56	110	60

The Krakatau eruption gave rise primarily to new surfaces of pumice and ash. Similar eruptuons in New Guinea (Taylor, 1957) produced vulcanoseres which also led rapidly to forest formation. Comparison of three areas devastated at different times gave the data shown in *Table 6.1*. The total of 110 species on Mt Victory after 87 years seems impressive, but the original forest of the area probably contained well over 500 species. White (1975) draws attention to the lengthy retention of secondary species, and points out that the 60 year old vulcanosere on Mt Bango in the Cape Hoskins area of New Guinea bears a forest which is structurally mature but floristically simple.

Screes of ash and coarser material are common on the Javan volcanoes where pioneer taxa include Gramineae, *Carex baccans*, Ericaceae, *Polygonum chinense*, *Anaphalis*, *Myrica* and ferns such as *Histiopteris* (van Steenis, 1972).

A longer record than these is obtainable by pollen analysis of crater lakes. In New Guinea, Lake Birip, at 1900 m, lies in a small explosion crater of andesitic ash about 2000 years old. Deposition in the lake (*Figure 6.3*) began with an inorganic clay containing little pollen. This presumably reflects a time when the freshly formed slopes around the lake were still unstable, and were insufficiently vegetated to contribute much pollen. This was succeeded by a phase of organic deposition which still continues, leaving a richly polleniferous sediment. An alternative explanation of the changeover in deposition would be that the lake only filled with water at that time (about 2000 B.P.). There is, however, no evidence in other pollen diagrams from the area (Sirunki and Inim, *Figure 5.16 and 5.18*) of a change to wetter conditions then. The pollen diagram (*Figure 7.7*) is in line with the seral hypothesis. The lowest sample contains very little pollen at all, and what is there is largely *Nothofagus* pollen which was almost certainly carried by wind from outside the volcanically devastated area. There is then a build-up of grass pollen, the peak of which coincides with the change from inorganic to organic deposition, and probably represents the development of grass-cover inside the crater. This is followed by a peak of tree fern spores, suggesting a possible phase of tree fern dominance. Subsequently mixed tree pollen is abundant (especially *Castanopsis/Lithocarpus* and *Macaranga*. The role of *Castanopsis acuminatissima* in New Guinea vegetation is still uncertain — it may be present in primary forest but almost pure stands are suspected of being secondary. *Lithocarpus* spp. are probably a primary dominant in forest at that altitude. The rate of this vulcanosere cannot be established clearly from the radiocarbon dates, but it may be estimated that *Castanopsis/Lithocarpus* forest (with *Nothofagus* present) was established in a few hundred years at most. The upper part of the diagram demonstrates forest clearance by man, which will be discussed in the next chapter.

The re-vegetation of lava streams seems to be slower than that of ash. On Mt Guntur in East Java, the lava flow of 1840 still shows as an almost bare black boot-shaped scab on the mountainside (van Steenis, 1972).

Even the much older lava flows nearby bear only sparse vegetation. Colonisation of lava streams in Bali was studied by de Voogd (1940), who found 40 species of plants on the lava stream of 1849, against only 10 (mostly ferns) on the stream of 1926. These values are considerably lower than those given above for volcanic ash.

Volcanic ash is frequently unstable and mud flows, known as lahars, often occur. These are really a special type of landslip, and the two may be considered together. Landslips are probably of some importance in primary forest in providing the areas which shade-intolerant pioneer species need to maintain themselves.

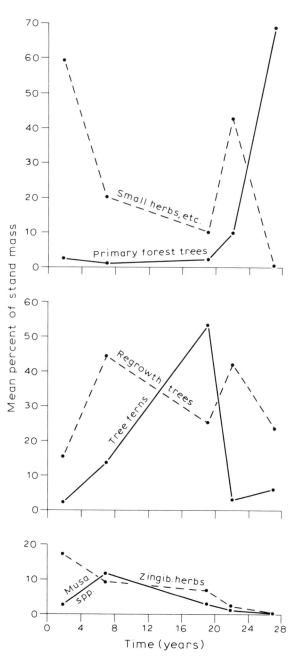

Figure 6.4 Changes through time in secondary vegetation in Mindanao (Philippines). (After Kellman, 1970)

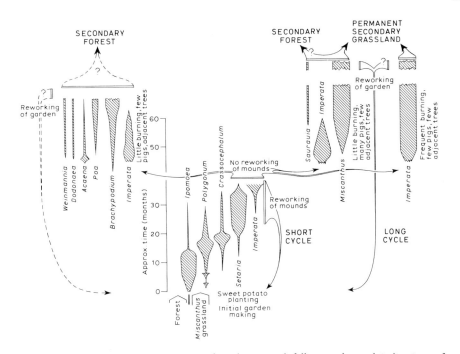

Figure 6.5 Schematic representation of gardening and fallow cycles and indications of possible regeneration series in a part of the New Guinea Highlands. The horizontal widths of the columns are intended to indicate the approximate relative abundances of the taxa they represent. (After Walker, 1966)

This, at least, is one tenet of the 'biological nomad' theory of van Steenis (1958a).

When trees do dominate these areas, which they usually do rather rapidly, several different species may be involved or one or two species may dominate. Examples of single-species dominance are the occurrence of *Casuarina junghuhniana* in East Java and *Pinus merkusii* in North Sumatra (van Steenis, 1972). On Mt Kelut in Java, the dominant pioneer trees were *Trema orientalis* and *Parasponia parviflora* (Clason, 1935). Successions on landslips are frequently similar to those following destruction of vegetation by human disturbance (see below).

The effect of cyclones on vegetation is very frequently highly destructive. Defoliation and breakage of branches represent minor damage from which trees readily recover. In severe conditions, however, destruction of whole trees in the forest may be total (Whitmore, 1974). After cyclone damage, light-demanding species enter the community temporarily (Whitmore, 1975). Areas where cyclone damage is a regular event appear to have distinct forest types. On Kolombangara in the Solomon Islands, Whitmore (1974) found that canopy disturbance, presumably by cyclones, was even more significant than altitude as a factor determining floristic composition.

The xeroseral type most commonly seen is undoubtedly the succession after forest destruction by man. Although it cannot be doubted (Whitmore, 1975) that when a cleared area is abandoned it can be occupied by one or more types of 'regrowth' culminating in 'secondary forest', the exact details of these events, and their rates and causes, have rarely been investigated. Mere arrangement of sample areas in a sequence is based on circular argument, and to establish what

actually happens it is necessary to have either permanent plots or at least plots of known history. Palynological evidence is rarely useful, for the wide source area which provides pollen for deposition at a single point is almost certain to contain many successional stages, whose record thus becomes confused.

Valuable studies of regrowth plots were carried out in West Malaysia by Holttum (1954), Kochummen (1966) and Wyatt-Smith (1966), revealing the course taken by successions of various types.

A good example of a study in which plots of known agricultural history were investigated is that by Kellman (1970) in the Philippines, who found that two distinct processes were at work (*Figure 6.4*). Many 'secondary' species established themselves shortly after abandonment. Owing to differing growth rates of these species, however, there was an *apparent* succession, as firstly herbs, then tree ferns and/or regrowth trees such as *Trema* spp. came to dominate the stand. After establishment of the canopy, and improvement in soil fertility, conditions were suitable for the second process, the re-establishment of the primary forest trees. The speed with which this took place, however, was related to availability of seed parents nearby, and Kellman estimated that several hundred years could be required for return to a fully diverse forest. It should be appreciated that Kellman's sites provide a series of glimpses into a number of slightly different successions, rather than a complete account of one succession, which could only be made from permanent quadrats.

The same problem confronted Greig-Smith (1952) in his analysis of secondary vegetation in Trinidad, and he concluded that two, possibly quite dissimilar, successions were taking place. One of these followed clearing of the forest, and the other succeeded clearing plus cultivation. The latter pursued a very variable course, but the former was more uniform, often involving temporary domination by the regrowth trees *Cecropia peltata* or *Ochroma pyramidale*.

Regrowth in the Nigerian rain forest areas (Ross, 1954; Jones, 1956) may begin with several small primary tree colonists including *Trema guineense* and *Vernonia conferta*. The former species is known to follow forest clearance without cultivation, but precisely which species dominates at first may be largely a matter of seed availability. In the shade of these early colonisers seed of the tree *Musanga cecropioides* germinates, and this usually dominates the succession from about 3 to 15 or 20 years. By this time many other species are established and a mixed rain forest develops. In fact Ross found that his 14- and 17-year old plots were as rich in tree species as the 'primary' forest in the same area enumerated by Richards (1939). One possible explanation of this is that much of the forest previously thought to be primary is in fact old secondary forest (Richards, 1955).

Studies of tropical regrowth by permanent quadrats are unfortunately very few; it is hoped the experiments started by Anderson (1960) and Schulz (1960) will be long continued. Schulz's experiments have already drawn attention to an important distinction between small clearings where light intensity is less than 10—20 times greater than in the forest, and larger openings with very high light intensities. In the former, the existing regeneration of the primary forest species made good growth, but in the latter the regeneration of these was strongly inhibited by the development of the secondary forest trees. A recent analysis of regrowth in permanent quadrats in sub-tropical rain forest in Australia (Williams *et al.*, 1969; Webb *et al.*, 1972) demonstrated a discontinuity (in time) between 'pioneer' and 'building' phases, reminiscent of the change at about 3 years in the Nigerian regrowth, and similar changes in Trinidad and the Philippines.

All these successions tend to lead back to forests, but in some areas, particularly those where there is a more pronounced dry season, or where human interference is more severe, the succession may be of a regressive type. Usually this leads to the establishment of a fairly permanent grassland which is maintained by frequent (often annual) firing (Bartlett, 1956). An example of the situation where the balance may be easily tipped either towards secondary forest or to grassland is given by Walker (1966) for the New Guinea Highlands (*Figure 6.5*).

6.3 HYDROSERES AND RELATED SUCCESSIONS

6.3.1 MANGROVES

Actual stratigraphic successions in mangrove have been demonstrated in northern South America by van der Hammen (1963) and Wijmstra (1969, 1971), as mentioned in Chapter 4 (*Figure 4.20* and *4.21*). The

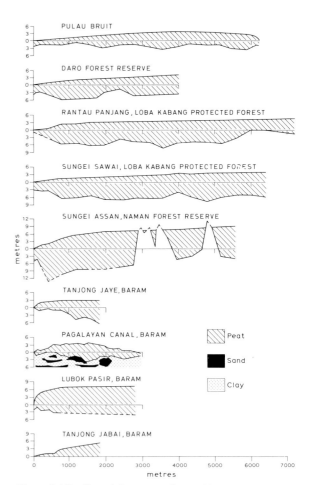

Figure 6.6 Profiles of the peat surface and base in peat swamps of Sarawak, Borneo. The domed surface is particularly clear. (After Anderson, 1964)

interpretation of seral changes is, however, complex, because the deposits probably reflect eustatic fluctuation of sea level as well as coastal accretion. In addition, since successive low 'glacial' sea levels occur in stratigraphic sequence, a gradual subsidence of the coastal area has probably occurred. The successions shown in *Figures 4.20 and 4.21* should be interpreted in the knowledge that *Avicennia* pollen has little mobility, whereas *Rhizophora* pollen is almost ubiquitous. The presence of *Avicennia* and *Rhizophora* pollen therefore probably indicates an *Avicennia* community, while a strong dominance by *Rhizophora* pollen alone may indicate a *Rhizophora* community (Muller, 1959). At present *Rhizophora* communities occur low on the beach and *Avicennia* higher up according to van der Hammen (1974), although the reverse is reported by Vann (1969) in at least some areas. The Late Pleistocene and Holocene succession in all cases is from dry land communities to *Avicennia*, to *Rhizophora* and back to *Avicennia*. The most likely explanation of this is that the first part, dry land communities to *Avicennia* to *Rhizophora*, represents the Late Pleistocene and the early Holocene eustatic rise of sea levels, while the subsequent change from *Rhizophora* to *Avicennia* represents the vegetational change consequent upon the build-up of deposits during the later Holocene phase of more or less steady sea level. This would confirm that *Avicennia* follows *Rhizophora* as the surface builds up, given constant land/sea relative level. Without more dates it is impossible, however, to be certain that this is the explanation.

6.3.2 PEAT SWAMP SUCCESSIONS

For many years the difficulties of working in these exceptionally tangled, humid and insect-rich environments deterred would-be investigators. In recent years, however, the indefatigable J.A.R. Anderson has made detailed studies of the finest peat swamps in the world in Sarawak, East Malaysia (Anderson, 1964). Stratigraphic profiles extending for more than 27 km show the domed shape of these swamps (*Figure 6.6*), which are analogous to the raised bogs of temperate regions. Pollen is excellently preserved in the peats and basal clays of these swamps, and many taxa have been identified. The vegetation is divided into six communities (which tend to occur roughly concentrically), by Anderson and Muller (1975), and the major palynologically identified arboreal taxa from these are shown in *Figure 6.7*. A small number of modern pollen rain (surface) samples suggests that these communities, as well as mangrove swamp, may be mutually distinguishable by their pollen rain (*Figure 6.8*). A core from community VI in the centre of Marudi peat swamp yielded 11.5 m of peat overlying a humic clay, which was bored to 13 m total depth. The pollen diagram (*Figure 6.8*) confirmed that the clay was a mangrove deposit, for the spectra from the base are rich in pollen of *Rhizophora*, *Nypa* and *Oncosperma*. Above 11 m there is a sharp change as these elements virtually disappear and *Campnosperma*, *Cyrtostachys*, *Palaquium*-type and *Dactylocladus* show a marked increase. This almost certainly indicates the presence of the *Campnosperma–Cyrtostachys–Salacca* subassociation, which normally occurs on shallow peat transitional from mangrove to peat swamp. The interval 8.5–6.5 m shows a rather sharp alternation in pollen spectra. Successively, these are dominated by *Gonystylus*, Myrtaceae, *Campnosperma* etc. and *Garcinia cuspidata*. These probably represent the occurrence of community I although the dominance of *Garcinia cuspidata* in the pollen spectrum is unusual. Between 6.5–6 m another major change occurs in the diagram, more or less coinciding with a change to less humified peat. *Dactylocladus*, in particular, increases here and a little higher, at 5 m, *Shorea albida*-type does the same. These samples up to 3 m depth, probably represent communities II and III. The top 3 m probably represents communities IV to VI, although the exact interpretation of the diagram is difficult. In general, however, this suggests that peat swamps can start growth on old mangrove swamps, and that the present vegetational zonation is probably a rough indication of the successional stages. It is interesting to note from the radiocarbon dates that the succession has taken approximately 4000 years. Presumably the mangrove swamp was able to form only after the eustatic rise of sea level, and 'living' peat swamps of much greater age are unlikely to be discovered except where tectonic movement has occurred.

The succession shows strong similarities to one from a Miocene brown coal (Anderson and Muller, 1975); indeed many of the pollen taxa are identical. It therefore seems clear that mangrove and peat swamp vegetation has, despite multiple changes in sea level, managed to maintain itself for several million years.

The succession from mangrove to peat swamps was also recorded in sediments from lowland Panama by Bartlett and Barghoorn (1973).

6.3.3 FRESHWATER SUCCESSIONS

Two of the major controls on the type of hydrosere which occurs in equatorial regions appear to be trophic status of the water (eutrophic or oligotrophic) and altitude. In general, as in temperate regions, early stages (after the open water phase) are dominated by herbs and later stages by woody plants, and the development of swamp forest is particularly strong in the tropics. Development of floating herbaceous swamp, analogous to the *schwingmoor* of temperate regions, seems also to be rather common. Floating islands of vegetation may even be present, as in the 'sudd' of the Nile (Walter, 1973), and in the Rawa Pening in Java (Polak, 1951).

The actual stratigraphic evidence of seral changes is not abundant, but a few examples will be described from well-developed hydroseres of different ecological types. To take first a eutrophic lowland swamp we shall consider the Rawa Lakbok in Java (Polak, 1949). The general stratigraphy is shown in *Figure 6.9* and detailed profiles are given in *Figure 6.10*. The stratigraphy shows that a volcanic ash layer stretches right across the swamp, although it is of rather variable thickness. The detailed profiles show that in most cases the deposition of the ash did not immediately affect the succession, for the deposit above the ash was similar to that below. The deepest profile, V, shows a first organic deposit described as sedge and fern peat with some forest peat (presumably wood fragments) particularly at the base. This is succeeded by susum peat (formed from the large subaquatic monocot, *Hanguana malayana*), with some forest peat at first, which continues to the surface except where interrupted by the volcanic ash. This suggests that forest peat has become less common as the hydrosere progressed — the reverse of the expected situation. This in fact is the case in four of the five profiles. The explanation is to be sought partly in the origin of the swamp which is a valley swamp, possibly formed by drowning of an originally forested valley. Another important fact is that lowland Java is almost entirely deforested now. Most remaining forest has for long been under human pressure. This is almost certainly responsible for the removal of swamp forest towards the surface of the deposit. Closer inspection of profiles III, IV and V shows that forest peat occurs in two

108

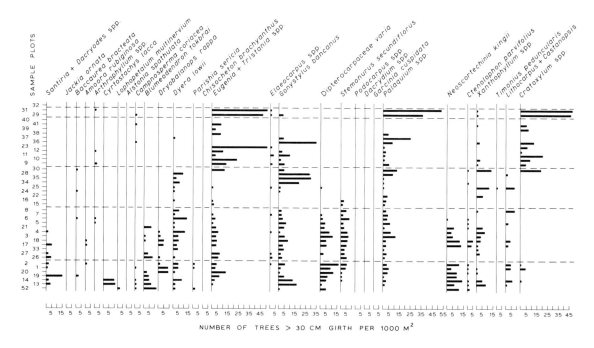

Figure 6.7 Present vegetation of a Sarawak (Borneo) peat swamp, showing the composition of the six major communities. (After Anderson and Muller, 1975)

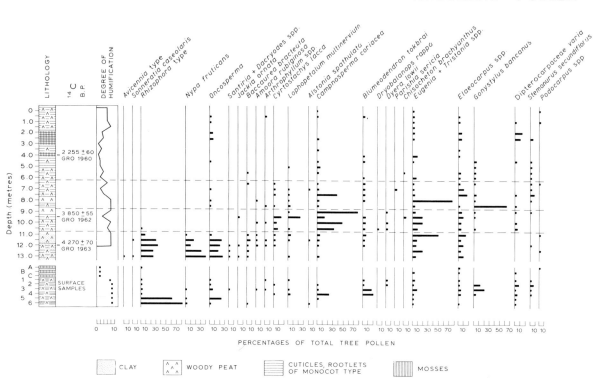

Figure 6.8 Modern pollen rain (below) and fossil pollen diagram (above) from Marudi peat swamp. Sarawak, Borneo. Only selected pollen types are shown. Values are percentages of total tree pollen. (After Anderson and Muller, 1975)

IN THE CATENA

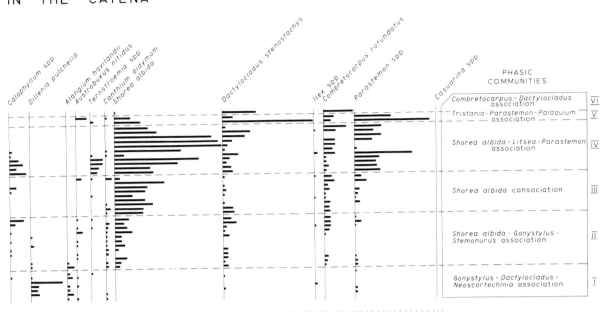

NUMBER OF TREES > 30 CM GIRTH PER 1000 M^2

IN PERCENTAGE

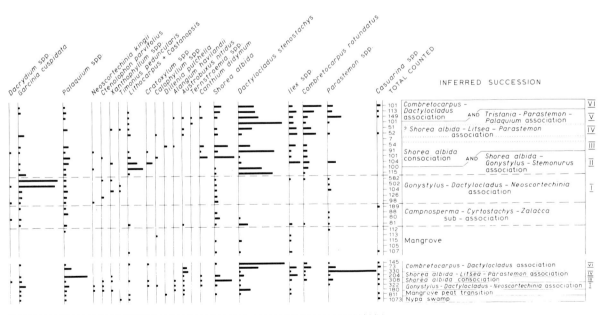

PERCENTAGES OF TOTAL TREE POLLEN

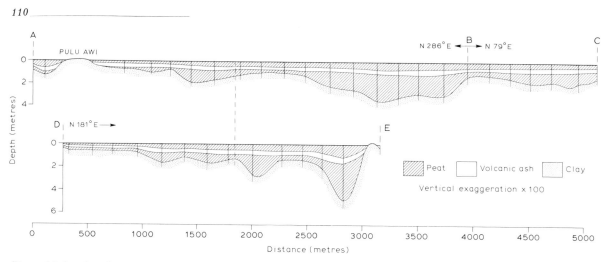

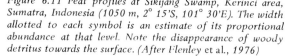

Figure 6.9 Stratigraphic sections through the peat deposits at Rawa Lakbok, Java, Indonesia. Vertical exaggeration 100×. (After Polak, 1949)

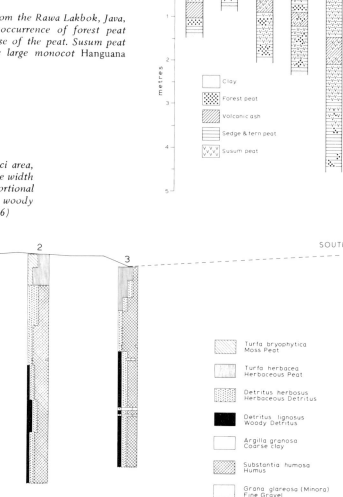

Figure 6.10 Individual profiles from the Rawa Lakbok, Java, Indonesia. Note especially the occurrence of forest peat (i.e. containing wood) at the base of the peat. Susum peat is composed of remains of the large monocot Hanguana malayana. (After Polak, 1949)

Figure 6.11 Peat profiles at Sikijang Swamp, Kerinci area, Sumatra, Indonesia (1050 m, 2° 15'S, 101° 30'E). The width allotted to each symbol is an estimate of its proportional abundance at that level. Note the disappearance of woody detritus towards the surface. (After Flenley et al., 1976)

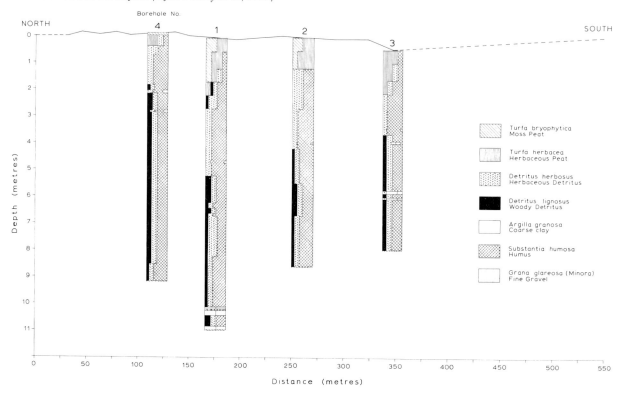

main layers. First there is a layer of forest peat — presumably the original drowned forest. Then comes a layer of sedge and fern peat (in profile III with susum peat and in profile IV with clay), which represents herbaceous swamp; then comes a new layer of forest peat — this time presumably swamp forest. Towards the surface this disappears and is replaced by a mixture of sedge and fern with susum peat, probably after the destruction of the swamp forest by man.

Examples of lowland swamp of oligotrophic type whose stratigraphy has been studied are very few (except for the coastal peat swamps already discussed) and we take an example from 1050 m altitude in Sumatra. Sikijang Swamp lies in a triangular basin that results from faulting. The swamp bears a fine vegetation of shrubs (including *Rhododendron javanicum* and *Vaccinium* sp.), growing from a soft carpet of *Sphagnum* in which five species of *Nepenthes* flourish (Flenley, unpublished observations; Morley *et al.*, 1973).

In the deepest borehole (*Figure 6.11*), the earliest organic deposit contains sufficient wood to be possibly a swamp forest deposit or the original dry land forest which preceded swamp formation. After a layer of clay (possibly weathered volcanic ash) the deposit is less woody but probably still indicates a swamp forest. Higher up wood is again abundant, clearly signifying swamp forest. The rather sudden disappearance of this wood from the succession is marked by fragments of charcoal and almost certainly this indicates destruction, by burning, of a swamp forest. In the absence of volcanic ejecta at this level it seems likely that the fire was lit by man, although lightning cannot be ruled out. Certainly the woody element of the present swamp vegetation is burnt readily enough by the local people, as I found to my cost while investigating the swamp. The later stages of this succession are therefore similar to those at Rawa Lakbok, except that *Sphagnum* rather than sedges and ferns dominates the post-clearance phase. The absence of an early treeless phase from the hydrosere may be because herbaceous and woody vegetation established simultaneously, or may be an artefact of the inadequate number of borings. Certainly treeless phases are present in the Danau Bento, a large swamp, now chiefly covered in swamp forest, at 1300 m altitude not far from Sikijang Swamp (Flenley, unpublished observations; Morley *et al.*, 1973). Indeed Danau Bento is entirely founded on a thick deposit of diatomite (*Figure 6.12*) which only forms in open water and is doing so abundantly in other Indonesian lakes (Hummel, 1931). Above the diatomite (diatom gyttja) there is a change either to herbaceous swamp or, as evidenced in at least two boreholes, directly to swamp forest. It must be remembered, however, that the wood in the deposit might have been tree roots penetrating a deposit of herbaceous detritus from above. Nevertheless, swamp forest certainly established, and is still abundant. A lens of water exists to the present day in the centre of the former lake, with a floating mat of grasses etc. above it.

For examples of sites at a rather higher altitude, we may turn either to Africa or New Guinea. In

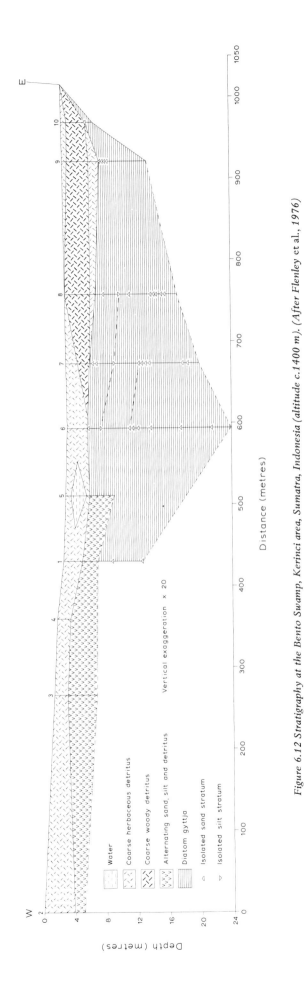

Figure 6.12 Stratigraphy at the Bento Swamp, Kerinci area, Sumatra, Indonesia (altitude c.1400 m). (After Flenley et al., 1976)

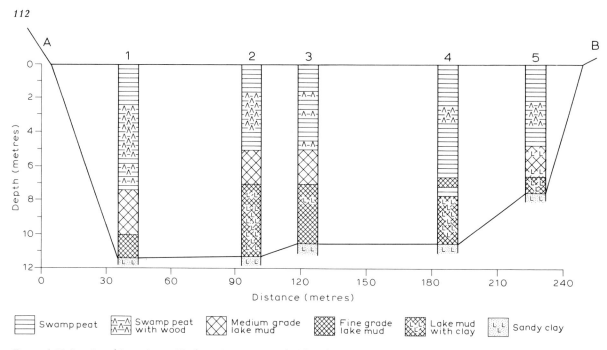

Figure 6.13 Stratigraphic section at Muchoya Swamp, Uganda (altitude 2256 m). Note especially the disappearance of wood from the peat near the surface. (After Morrison, 1968)

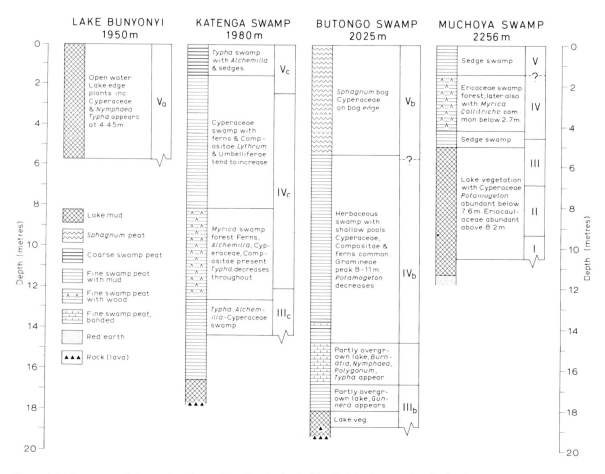

Figure 6.14 Summary of the stratigraphy at four sites in the Rukiga Highlands, Uganda. The local vegetation, as reconstructed from the pollen record, is also given and the pollen assemblage zones recognised in each core are indicated. (After Morrison and Hamilton, 1974)

Uganda, Muchoya Swamp, at 2260 m, which furnished us with an important pollen diagram for Chapter 3, has an interesting stratigraphy (*Figure 6.13*) (Morrison, 1968). The site clearly started as a lake originating possibly by uplifting of a valley. Soils in the area are generally poor and acidic (Morrison and Hamilton, 1974), so the site is probably oligotrophic or mesotrophic. In most boreholes the original fine detritus mud passes upwards into a coarser detritus mud, and thence to a swamp peat which sooner or later contains an abundance of wood. This seems like a typical hydrosere, in which open water gives way to herbaceous swamp which is replaced by swamp forest. In all boreholes evidence of this swamp forest disappears near the surface, to be replaced by herbaceous swamp deposits again. The present swamp is dominated by the tussock-forming sedge *Pycreus* sp. A pollen diagram showing the aquatic and likely swamp elements largely confirms this. The early phases are dominated by pollen of *Potamogeton*, an open water plant. A rise of Cyperaceae pollen is closely followed by a change to coarser detritus mud. The appearance of wood coincides approximately with rises in Ericaceae and swamp trees. *Myrica* pollen eventually reaches very high values. Morrison postulates a development of *Erica rugegensis*, followed by a swamp forest dominated by *Myrica kandtiana*. Both of these species occur on the swamp today. The disappearance of the swamp forest is mysterious. Where man has felled or burnt swamp forest elsewhere in Uganda, it has been replaced not by *Pycreus*, but by *Cyperus papyrus* or *Cyperus latifolius*. Morrison postulates a change either in drain-

age of the valley, or in climate, putting back the hydrosere by several thousand years.

The stratigraphy of three other swamps from the same area is compared with that of Muchoya in *Figure 6.14* (Morrison and Hamilton, 1974). Lake Bunyonyi's hydrosere is clearly in an early stage of development. That of Katenga Swamp at 1980 m is rather similar to Muchoya, but the upper sedge (Cyperaceae) swamp gives way to *Typha* swamp near the surface. The reasons for this change are unknown, but it coincides with other changes, possibly reflecting forest clearance round about. The authors therefore suggest that 'Perhaps forest clearance encouraged soil erosion and increased quantities of sediment and of cations reaching the swamp surface; conditions favourable for the growth of these plants were thereby created'. Butongo Swamp at 2025 m presents a considerable contrast with Muchoya and Katenga. There is no evidence of swamp forest at this site: instead the herbaceous swamp gives way to *Sphagnum* bog, which has persisted to the present day. Butongo Swamp is in an area of higher rainfall (Morrison, 1968) and this may be one contributary cause for the difference. An alternative possibility is that the Butongo water is particularly acid and oligotrophic, but this would be surprising in view of the fact that the site is dammed by a lava flow, the weathering of which might be expected to be releasing nutrients into the swamp.

Although the Katenga core is unfortunately not dated, the pollen diagram (*Figure 7.2*) suggests that development of swamp forest there and at Muchoya occurred at roughly the same time. This also coincides

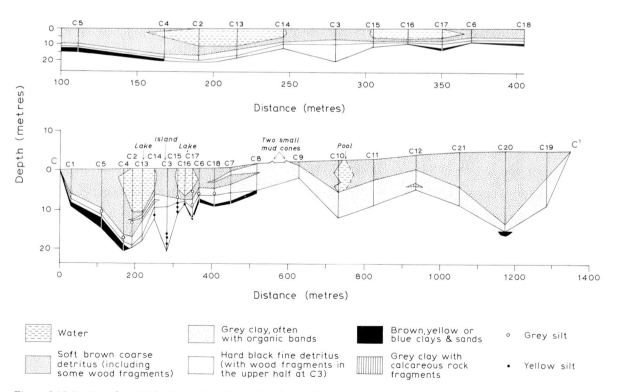

Figure 6.15 Stratigraphy at Lake Inim, New Guinea Highlands (altitude 2550 m). Below is the complete section with a vertical exaggeration of 10X. Above is the section through the lake without vertical exaggeration. (After Flenley, 1967; Walker and Flenley, in press)

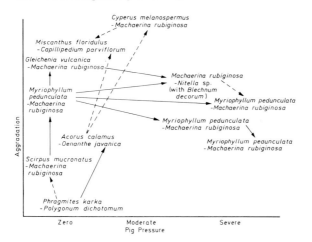

Figure 6.16 Hypothetical courses of the hydrosere in a part of Sirunki Swamp, New Guinea Highlands (altitude 2500 m). Continuous lines define courses for which there is some evidence. (After Walker, 1972)

with the change to banded sediment at Butongo (*Figure 6.14*). This synchroneity suggests a regional climatic change, presumably towards dryness. Thus this apparently 'normal' hydrosere may be the result of environmental change, rather than autogenic processes.

Stratigraphy has been carefully recorded at a number of upland sites in New Guinea. Lake Inim, 2550 m altitude, is a mesotrophic lake with a pH of 7.5, because it comes under the influence of nearby limestone (Flenley, 1967, 1972). Nevertheless, rainfall is adequate to support *Sphagnum*, and this develops on floating mats where these build up above water level, thus keeping the plant away from the influence of the alkaline groundwater. The stratigraphy (*Figure 6.15*) reflects three main phases. An early open water phase led to the deposition of fine detritus mud which is now well compacted. Subsequently a mud volcano erupted, depositing a layer of grey clay at the down-valley end of the lake. The third phase was of coarser, softer material, chiefly remains of *Machaerina rubiginosa*, the sedge which dominates the lakeside swamp today. Near the surface of the swamp, timber was frequent in the marginal deposits, and tree stumps indicated the presence of former swamp forest, chiefly *Phyllocladus hypophyllus, Dacrycarpus imbricatus* and *Eugenia* sp. Local people averred they had cut down this forest, within living memory, for firewood. Thus, until interference by man, the hydrosere appears to have followed a fairly predictable course. A much more complicated situation appears to be present at the nearby Kayamanda (Sirunki) Swamp (Walker, 1972a); this is a much larger and more varied site with a history exceeding 40 000 years. There is no evidence of swamp forest except in marginal areas (R.G. Robbins, personal communication). Human influence on the site appears to be exerted not so much directly – although burning occurs – as by domesticated swine, which use parts of the site as a wallow, and rootle there for food. Possible courses of the hydrosere are shown in *Figure 6.16*, in which the continuous lines indicate courses for which there is some evidence (stratigraphic or otherwise).

As an example of a high altitude site we may take Imbuka Bog at 3550 m on Mt Wilhelm (Hope, 1976). The bog is dammed by a small Pleistocene moraine. It is well below the limit of closed forest, but only forest remnants survive nearby at the present day. The present bog surface is vegetated by short grass bog, *Deschampsia klossii* tussock grassland, sub-alpine *Astelia* bog and *Carpha alpina*- fen grassland. Hope records the following stratigraphy:

0–10 cm	Tussock bases, coarse fibrous peat.
10–50 cm	Deep brown amorphous peat with numerous rootlets.
50–231 cm	Deep brown peat with wood fragments, fibres, leaves and *Rapanea* fruits. Becoming increasingly firm with depth.
231–238 cm	Fibrous brown peat.
238–241 cm	Volcanic ash band.
241–397 cm	Brown amorphous algal gyttja with occasional plant debris including wood at 270 and 318 cm and monocot leaves and fibres.
397–399 cm	Bands of yellow clay.
399–440 cm	Brown to deep brown gyttja.
440–450 cm	Sticky yellow clay above stone basement (possibly moraine).

Volcanic ash bands were encountered at 30, 91–93, 238–241, 289–294, 400–405 and 412 cm. The radio-carbon age of a sample from 415–440 cm was 12 570 ± 290 years. The stratigraphy clearly suggests that deglaciation left a moraine-dammed pond which filled with algal gyttja until peat formation could occur. The earliest peat was 'fibrous' (i.e. presumably of herbaceous origin), but this was quickly succeeded by wood peat suggesting sub-alpine shrubbery or forest. This woody vegetation apparently reverted more recently to herbaceous bog vegetation. Hope concludes that this change was probably the result of human activity beginning about 800–1000 years ago. All these changes are confirmed by pollen evidence.

The successions which have now been demonstrated in the foregoing studies are summarised in a very generalised way in *Figure 6.17*. It will be clear from this diagram that equatorial hydroseres often do show a succession from open water through herbaceous swamp to swamp forest. It will be equally clear that several variations on this theme are possible. Firstly, not all swamps start as open water. Marudi peat swamp developed over mangrove; Rawa Lakbok and Sikijang Swamp appear to have developed directly by the swamping of an existing forest without any open water stage. Secondly, not all herbaceous swamps turn over to swamp forest, the particular exception in our examples being Butongo Swamp, which changed to *Sphagnum* bog. Thirdly the change to swamp forest may not always be strictly autogenic; at Katenga and Muchoya Swamps the synchroneity of the change suggested that a climatic drying was responsible. Fourthly, many sites show the later disappearance of swamp forest. In New Guinea and Sumatra this appears to be due to felling and/or burning, but in the Uganda

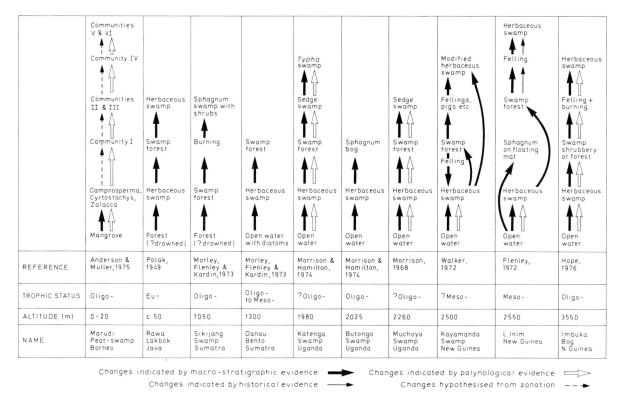

REFERENCE	Anderson & Muller, 1975	Polak, 1949	Morley, Flenley & Kardin, 1973	Morley, Flenley & Kardin, 1973	Morrison & Hamilton, 1974	Morrison & Hamilton, 1974	Morrison, 1968	Walker, 1972	Flenley, 1972	Hope, 1976
TROPHIC STATUS	Oligo-	Eu-	Oligo-	Oligo- to Meso-	?Oligo-	Oligo-	?Oligo-	?Meso-	Meso-	Oligo-
ALTITUDE (m)	0-20	c.50	1050	1300	1980	2025	2260	2500	2550	3550
NAME	Marudi Peat-swamp Borneo	Rawa Lakbok Java	Sikijang Swamp Sumatra	Danau Bento Sumatra	Katenga Swamp Uganda	Butongo Swamp Uganda	Muchoya Swamp Uganda	Kayamanda Swamp New Guinea	L.Inim New Guinea	Imbuka Bog N Guinea

Changes indicated by macro-stratigraphic evidence ➡   Changes indicated by palynological evidence ⇨
Changes indicated by historical evidence ──➤   Changes hypothesised from zonation ─ ─➤

Figure 6.17 Summary of changes recorded in the equatorial hydroseres mentioned in the text. (Original)

sites the cause was uncertain. There is also evidence that man may affect the type of herbaceous swamp; for instance, at Katenga, sedge swamp probably changed to *Typha* swamp under human influence, and at Kayamanda several changes were thought to be due to domestic pigs. It is surely true that these few examples by no means represent the full range of variation, and that any complete account of equatorial hydroseres is far in the future. It is hoped, however, that enough has been said to demonstrate the value of stratigraphic and palynological studies in this field of research.

6.4   CONCLUSIONS

The aim of this chapter has been to show that successions in equatorial regions are very variable, and

thus fail to fit adequately to the old 'climax' theories of succession. Since the concept of the Clementsian 'climatic climax' has, like that of the Davisian cycle of erosion, been severely modified in modern ecological thinking (e.g. Whittaker, 1953; Raup, 1964), ideas on successions have become less rigid than formerly. This is particularly true in temperate regions where climatic change is known to have been extreme in the past. In equatorial regions, vegetational change related to climatic variation has now been shown (Chapters 3–5) to have been considerable, although perhaps less marked than in temperate regions. Possibly, therefore, one might think that ideas of successions could still be fairly rigid in the equatorial regions. There is another way, however, in which it may turn out that the equatorial regions exceed the temperate in their variability with time: in the effects of man on vegetation. This is the subject of the next chapter.

7
The Influence of Man

7.1 INTRODUCTION

The severity of man's attack on vegetation has been as great in equatorial regions as in many other parts of the world, and the survival of even semi-natural vegetation in many areas is a tribute to the capacity of the vegetation to resist and regenerate rather than to human self-restraint. Although he must always have been affecting vegetation, most of the effects of very early man cannot be detected. An exception here is the case of fire, which may have been used in very early times for hunting, especially outside perhumid areas. Unfortunately it is difficult to distinguish between natural fires and those lit by man. One desirable line of research would be to establish when the frequency of fires first increased beyond the 'background' rate. This might be possible by measuring the quantity of charcoal in ocean cores, as suggested by Smith, D.M. et al. (1973).

Direct evidence regarding the influence of more recent man on vegetation is of two main kinds: palynological indications of vegetational changes such as forest clearance, and archaeological records of the plants utilised, or in some cases domesticated, by man. It is assumed that evidence of cultivation of domesticated plants is evidence of modification of natural vegetation, since almost all known cultivation practices (especially primitive ones) involve forest clearance (Figure 7.1) or some other major disturbance. There is also much data from plant geography and cytogenetics which is of importance – but most of that will not be covered here because it is indirect evidence.

Since the agricultural activities, and particularly the crops, have been somewhat different in the three main equatorial regions, we shall revert here to a regional treatment.

7.2 EQUATORIAL AFRICA

It is appropriate to consider Equatorial Africa first, since the classical discoveries at Olduvai Gorge, Lake Rudolph and elsewhere have established the presence in East Africa of *Homo erectus* and other primitive forms of man in exceptionally early times. Indeed Africa is now commonly claimed as the evolutionary home of mankind (Livingstone, 1971a); this may well be true and the idea is supported by the abundance of human parasites in Africa and by the presence of man's nearest relative, the chimpanzee, but one must be wary of premature judgements based on negative evidence from other areas. Modern man, *Homo sapiens*, was present in Africa by c.60 000 B.P. (Clark, 1970). Unfortunately we have little early evidence of man's influencing vegetation, although the savannah vegetation around Lake Victoria before 12 000 B.P. (Kendall, 1969) would probably have been rather easy to fire for hunting.

Early evidence of agriculture is also slight but this may be because we have so far been looking in the wrong place. The indirect evidence of plant distribution (Vavilov, 1951) suggests that the chief centre of plant domestication in equatorial Africa was in Ethiopia where about 38 species were possibly first cultivated

Figure 7.1 Forest clearance in Amazonia. A recently felled area in lowland rain forest near Iquitos, Peru. (Photograph by C. D. Ollier)

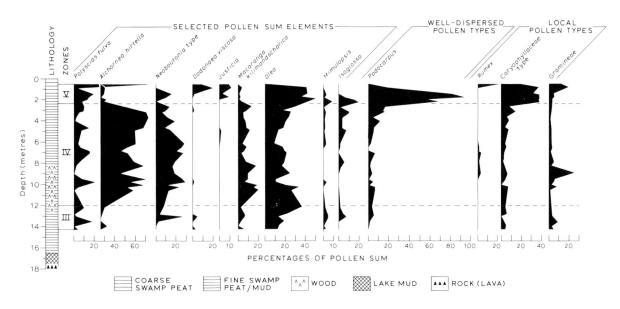

Figure 7.2 Pollen diagram from Katenga Swamp, Rukiga Highlands, Uganda (altitude 1980 m). Only selected taxa are shown. Values are percentages of a pollen sum containing those taxa thought to have originated from plants growing close to (but not on) the swamp. (After Morrison and Hamilton, 1974)

or used. These include important species such as *Triticum* spp. (some of the wheats), *Hordeum sativum* (barley) and *Coffea arabica* (coffee), which have since spread to many other areas. It is therefore to be expected that Ethiopia may eventually yield the earliest evidence of agriculture in Africa. There are, however, pollen diagrams from Uganda which show probable forest clearance from c.1000 B.P. One of the most interesting is that from Katenga Swamp at 1980 m (Morrison and Hamilton, 1974), shown in *Figure 7.2*. Most of this diagram is occupied by a Zone designated IVc in which pollen types such as *Alchornea, Macaranga, Neoboutonia*-type, *Olea* and *Polyscias fulva* are abundant. All of these suggest a forest strongly resembling the *Prunus* zone of the present day forests of the area. At the beginning of the succeeding Zone Vc the percentages of many of these, particularly *Alchornea* and *Macaranga*, are much reduced. At almost the same level there are rises in the values for Caryophyllaceae-type and at a slightly higher level *Justicia* and *Dodonaea viscosa* become significant for the first time. Higher still there are peak values for Gramineae and *Rumex*. Surface samples from cleared or partially cleared land in the Rukiga Highlands often show high pollen values for Caryophyllaceae-type, *Dodonaea*, Gramineae, *Justicia* and *Rumex* pollen (Hamilton, 1972). Zone Vc therefore strongly suggests that forest clearance was occurring on the dry land around the swamp. It is true that two tree-pollen types, *Olea* and *Podocarpus*, show high values in this zone, but as these are probably well or very well dispersed (*see* Chapter 3), their increases are probably artefacts reflecting merely a reduction in the total pollen reaching the swamp from nearby.

The possible course of events which would account for this part of the pollen diagram is described thus by Morrison and Hamilton (1974):

'*Stage 1*. Trees on lower slopes, such as *Alchornea hirtella*, were felled, allowing a temporary increase, either in abundance or in flower production, of understorey shrubs such as *Isoglossa* and *Mimulopsis*. *Neoboutonia macrocalyx* and *Polyscias fulva*, both of which can grow in secondary and primary forest in the Rukiga Highlands, remained in adjacent vegetation. Forests on the less fertile upper slopes were not affected by this clearance so that pollen types like *Olea* and *Podocarpus*, which are produced by upper slope trees and which are well dispersed, increased in abundance in the pollen diagram as adjacent pollen production declined.

Stage 2. Increased pressure on the land led to replacement of secondary forest trees such as *Neoboutonia* and of understorey shrubs such as *Isoglossa* and *Mimulopsis* by herbs and shrubs such as *Dodonaea* and *Justicia* which are characteristic of more intensively managed land. The rotation was thus shortened to exclude forest regeneration beyond the shrub stage. The replacement of the Acanthaceae shrubs, *Isoglossa* and *Mimulopsis*, by the shrub *Dodonaea* could be due to soil deteriora-

tion. The evidence of increased silting in the Katenga sedimentary basin makes such an explanation very likely.

Stage 3. Further increase in human disturbance resulted in destruction of upper slope forest trees such as *Olea* and *Podocarpus*, the latter apparently being felled first. The very high values of *Polyscias* pollen at the top of this stage are likely to be related to the generally low pollen production of adjacent vegetation so that isolated *Polyscias* trees growing close to the swamp are represented in the pollen sum out of all proportion to their actual abundance in the vegetation.'

The regional rather than local nature of these changes is attested to by the short diagram from Lake Bunyonyi which shows many similarities to that from Katenga, although the two coring sites are about 14 km apart. That similar changes were not seen in the diagram from Muchoya Swamp (*Figure 3.19*) is not surprising, for that is at a higher altitude, in an area not cleared of forest even today. Unfortunately, there are no radiocarbon dates from the Katenga site, but by correlation with the dated Muchoya sequence, it seems likely that the complete core covers about the last 8000 years. If sediment had accumulated at a steady rate (a large assumption), forest clearance would have begun about 1000 B.P. in the area. This date corresponds reasonably well, however, with the maximum age of c.1000 A.D. for the Dimple-based pottery industry of East Africa (Posnansky, 1967). It is believed that the makers of these ceramics were cultivators. Livingstone's (1967) pollen diagram from the Ruwenzori (*Figure 3.14*) also suggests forest clearance from about the same date on the plateau below Ruwenzori. The excellently dated Lake Victoria pollen diagram (*Figure 3.20*) shows a fall in forest pollen types (*Celtis* and Moraceae particularly) and a sharp percentage rise in Gramineae, from about 3000 B.P. These changes are consistent with forest clearance, although a climatic explanation cannot be ruled out. That clearance should start earlier in some places than others is scarcely surprising, and according to Clark (1962) agriculture was already established in East Africa by 3000 B.P.

It must not be concluded that all cultivators in tropical Africa were clearers of forest. In Rhodesia at 18° S, Tomlinson (1974) shows that large Gramineae pollen grains, possibly those of cultivated grasses, appear in the record at a time of reduction of other grasses and of Compositae pollen. All the earlier part of the pollen diagram shows abundant Gramineae pollen, and presumably when cultivators arrived they cleared savannah rather than forest (at an estimated date of c. 3000 B.P.).

7.3   EQUATORIAL LATIN AMERICA

The time of first arrival of man in the Americas is a highly contentious issue. On the one hand there are those (e.g. Haynes, 1974; Lynch, 1974) who argue

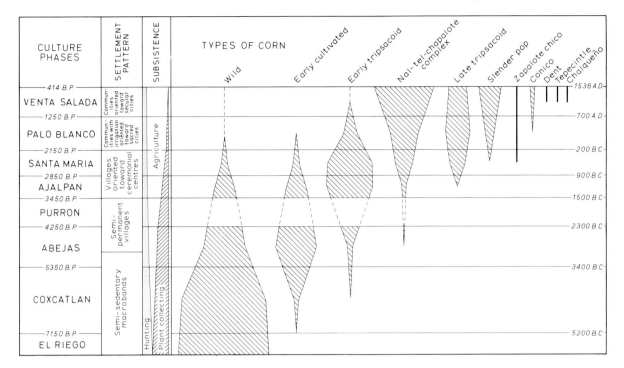

Figure 7.3 Frequency polygons, in terms of percentages of cobs identified, showing changes in the types of corn in the Valley of Tehucan, Mexico, from 5200 BC to AD 1536. Specimens of prehistoric corn were almost totally lacking for the Purron culture, which is recognised by other means. (After Mangelsdorf et al., 1964)

that the first men in the western hemisphere arrived on foot over the Bering Strait during the last glaciation low sea level time (possibly about 14 000 B.P.) and that the first arrivals were essentially big game hunters responsible for the extinction of many large mammals in North and South America (e.g. the mastodon and the giant ground sloth). Indeed it has even been suggested that a 'frontier' of hunters might have taken only 1000 years to move from Canada to Tierra del Fuego, overkilling on their way (Martin, 1973). On the other hand are those who point to a growing number of dated archaeological sites which are too early to fit into this scheme (Bryan, 1973). Unfortunately most of these early dates have been seriously questioned for one reason or another so that the time of arrival of man in the Americas must remain uncertain.

As far as vegetation is concerned, this controversy is not necessarily directly relevant, for people did not, so far as we are currently aware, seriously affect the vegetation until they began forest clearance for agriculture, although their hunting fires may have been of importance earlier. Fires were certainly occurring in the New World tropics in Late-glacial time (Kellman, 1975). There is no need to regard cultivation as anything but an indigenous development in the Americas, although there is no evidence which categorically excludes the possibility of very early trans-oceanic contacts, and some which is claimed to support the possibility (Riley *et al.*, 1971).

The geographical distribution of wild forms of the cultivated plants of the tropical Americas, and of wild species related to them, suggest that the cultivars fall into two groups, and that cultivation may have originated separately, therefore, in two main centres

(Vavilov, 1951). These are the South Mexican and Central American Centre and the South American (Peruvian—Ecuadorean—Bolivian) Centre.

The American domesticates include a number of important root crops, particularly sweet potato (*Ipomoea batatas*), cassava (*Manihot esculenta*) and potato (*Solanum* spp.). They also include the grain crop, maize (*Zea mais*). There has been a belief, fostered by Sauer (1952), that as a general principle root crops were domesticated first and grain crops later. This idea may be tested against the evidence given below. The origin of the three major root crops mentioned above, although not known in detail, is not really mysterious. All have close wild relatives from which it is likely they are derived.

The origin of maize is a rather more difficult problem. Its only wild relatives, teosinte (*Zea mexicana*) and *Tripsacum* are not at all similar to any maize variety, and there is no wild maize species. Alternative theories derive maize either from teosinte or *Tripsacum* via one or more massive mutations or from a now extinct hypothetical wild maize. The earliest cobs, excavated from caves in the Tehuacan valley, Mexico, dated back to c. 7150 B.P. and were regarded as wild corn by Mangelsdorf *et al.* (1964). The 'wild' cobs are believed to have borne both the 'pod' character of having the grains enclosed in glumes and the 'pop' character of possessing hard seeds which explode when heated. By crossing modern varieties with these characters, Mangelsdorf eventually produced cobs similar to the 'wild' corn cobs in the deposits. The subsequent development of more modern varieties at the Tehuacan valley site, and the hypothetical development of agriculture, is shown in *Figure 7.3*.

Palynological evidence has also been important regarding the origin of maize. Pollen of maize is distinct from that of teosinte, *Tripsacum* and other grasses by its exceptional size (c. 85–120 μm), morphological details and its appearance under phase-contrast illumination (Bartlett *et al.*, 1969; Irwin and Barghoorn, 1965; Tsukada and Rowley, 1964). Pollen of maize has been identified in deposits dating from c. 70 000 B.P. under Mexico City (Barghoorn *et al.*, 1954), at which time it was surely wild (*Figure 7.4*). Despite this apparently strong evidence for a now extinct 'wild' corn, Galinat (1971) has recently revived the idea of origin of maize directly from teosinte.

We shall consider the evidence for forest clearance and cultivation together, first for areas near the northern of Vavilov's centres (i.e. Central America) and then for those near the southern centre (i.e. South America). For Central America, we have already mentioned the evidence for Tehuacan. Tsukada (in Cowgill *et al.*, 1966) found maize-type pollen along with that of weeds in two cores from Laguna de Petenxil, Guatemala. The period covered was from about 3990 B.P. and maize pollen reached a peak about 1600 B.P.

Additional evidence of maize came from the cores obtained in the Gatun basin through which the Panama Canal runs (Bartlett *et al.*, 1969). At an age of c. 7000 B.P., partly supported by radiocarbon evidence, the first maize pollen grains were found. These were slightly smaller and thicker walled than those of cultivated maize and are presumed to originate from the wild ancestor. At a later period, approximately 3100 to 1800 B.P., there was pollen referable to cultivated maize, accompanied by high values for pollen of Gramineae and Compositae (*Ambrosia*-type), and of the herbaceous weed *Borreria*. There was also a scarcity of tree pollen; all these points combine to suggest forest clearance and agriculture. Later still, about 1800 B.P., there is an occurrence of pollen of *Manihot* and *Ipomoea*. In the latter case one cannot be certain of the species, nor whether it was wild or cultivated, but in the case of *Manihot* the pollen was recognisable as a cultivated cassava type (*Figure 7.4*).

In upland Panama, at 1800–2375 m, Linares *et al.* (1975) have found carbonised remains of maize and beans (*Phaseolus vulgaris*) dating back to 2380 ± 60 B.P. The same bean was, however, recorded at Tehuacan from 7000 B.P. (Smartt, 1969).

In addition to actual evidence of crop plants we have the important evidence of forest clearance. This has been mentioned above for the Gatun cores, and similar phases of high values for Chenopodiaceae/Amaranthaceae pollen are taken to indicate Aztec agriculture in Mexico at Lake Patzcuaro (Hutchinson *et al.*, 1956). The Laguna de Petenxil cores from Guatemala mentioned above also contain clear evidence of vegetational disturbance in Maya time, and this was confirmed at several other sites (Tsukada and Deevey, 1967).

The general conclusion to be drawn from Central American evidence is that maize and beans have been cultivated there since at least 7000 B.P. Root crops

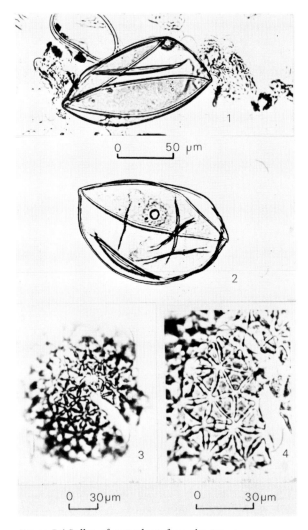

Figure 7.4 Pollen of crop plants from the Americas.
*1. Fossil maize (*Zea mais*) pollen from c.70 m depth below Mexico City. The age of this specimen is c.70 000 B.P.*
*2. Modern maize (*Zea mais*) pollen.*
*3. Fossil pollen of cassava (*Manihot esculenta*) from the Gatun Basin sediments, with an age of c.1800 B.P.*
4. Detail of 3, showing the surface pattern which distinguishes cultivated from wild Manihot *pollen. (After Barghoorn* et al.*, 1954 and Bartlett* et al.*, 1969)*

appear in the record only from 1800 B.P. and in the extreme south of the area (Panama). The record is of course very scanty, and later conclusions may be different.

There is now some archaeological evidence available from near Vavilov's southern centre of domestication. On the desert coast in Peru indications have been found of cultivation of the bottle gourd (*Lagenaria siceraria*) and a species of squash (*Cucurbita* sp.) in a phase dating from c.7200 B.P. to c.6200 B.P. (Patterson, 1971; Moseley, 1975). The people concerned appear to have relied chiefly on fishing and shellfish collecting, however. This cultivation continued through the ensuing phase (c.5700 to c.4500 B.P.), and subsequently became progressively more important and included other crops such as cotton (*Gossypium barbadense*), guava (*Psidium guajava*), jack beans (*Canavalia* sp.), chili peppers (*Capsicum* spp.), avocados

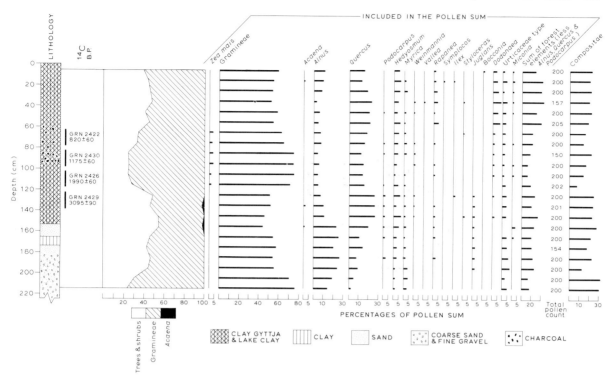

Figure 7.5 Pollen diagram from Laguna de los Bobos, Colombia, South America (altitude 3800 m). Only selected pollen types are shown. The percentages are based on a pollen sum including trees and shrubs, Gramineae and Acaena. (After van der Hammen, 1962)

(*Persea americana*) and coca (*Erythroxylon coca*). The earliest *Capsicum* grown was *C. baccatum* (Pickersgill, 1969). The bean *Phaseolus lunatus* is known from Chilca Peru, at 5300 B.P. (Smartt, 1969) and the groundnut *Arachis hypogaea* from Huaca Prieta, Peru, at 2500 B.P. (Krapovickas, 1969). Maize appears late and in small quantities in most Peruvian sites (Moseley, 1975).

The pollen diagram from Laguna de los Bobos at 3800 m in Colombia (van der Hammen, 1962) shows a close correlation between reduction in forest pollen and presence of maize pollen (*Figure 7.5*). The maize cultivation, which probably occurred in valleys well below the site, is shown by radiocarbon dates to have been from c. 2250 to c. 750 B.P., very approximately the time of the Inca civilization in South America.

In fact the majority of Holocene pollen diagrams from Colombia (*see* Chapter 4) indicate forest clearance in their upper parts. At La Fuquene (van Geel and van der Hammen, 1973) this was evidenced by rises in the values of *Dodonaea* and Gramineae accompanied by decline in the values of forest elements, beginning just after c. 3000 B.P. The behaviour of *Dodonaea* which is a shade-intolerant pioneer shrub and occurs in both the Late-glacial and the clearance phase, is reminiscent of that of *Artemisia* in Europe. A similar rise in Gramineae pollen values and decline of forest elements is evident near the top of the Sabana de Bogota diagram (van der Hammen and Gonzalez, 1960) and that from El Abra (van der Hammen, 1974).

Excavations in savannah areas of lowland Venezuela by Zucchi (1973) revealed a most interesting result. In an earlier phase (2870–1450 B.P.) maize cobs were

present, but after that time cassava appears to have replaced maize as the chief crop.

Three significant points emerge from this South American evidence. Firstly, the earliest crops were seed crops, not root crops. Secondly, maize occurs much later than in North America. Thirdly, at the Venezuelan site, cassava follows rather than precedes maize.

The question of whether root crops preceded or followed grain crops into cultivation in the Americas cannot yet be finally resolved, but in at least two areas (Panama and Venezuela) maize has been shown to have been grown well before cassava; the hypothesis of Sauer (1952) may therefore be provisionally discarded. An alternative idea, that maize and beans (as staples) came southwards from Mexico, while cassava came northwards from South America (Linares *et al.*, 1975), seems more likely but is as yet unsubstantiated.

7.4   EQUATORIAL INDO–MALESIA

Man, in the form of *Homo erectus*, inhabited Java in the Early Pleistocene (van Heekeren, 1972; Sartono, 1970; Jacob, 1973) possibly as early as 1·7 M years ago (Lestrel 1976) and modern man (*Homo sapiens*) was recorded at Niah cave in Borneo c. 40 000 years ago (Harrison, 1970). The aborigines of Australia had certainly arrived there by 32 000 B.P. (Barbetti and Allen, 1972) and may have been significant in diminishing the Australian mega-fauna about that time (Jones, 1975). There is no real evidence as yet that man was significantly affecting vegetation in these early times, although Kershaw (1975b) has suggested that burning by aboriginal man could have been at least partially responsible for replacement of an *Araucaria*-dominated rain forest by eucalypt vegetation between 38 000 and 27 000 B.P. in northern Queensland (*see* Chapter 5). Most of the natural vegetation of the whole region probably survived, however, until the inception of agriculture.

Vavilov (1951) argued that China, with its great number of domesticates (136 spp.) must have been the world's oldest centre of cultivation, but considered India, which is the home of rice and many other important crops, to be the most significant. South-East Asia was also considered a centre of some importance, and Sauer (1952) suggested it as the 'hearth' or place of origin of all agriculture, whence it had diffused all over the world.

Evidence of possible very early agriculture (before 15 000 B.P.) in southern India has been claimed by Gupta (1971), who found pollen grains, believed to be those of cereals, associated with *Artemisia* pollen and other weed types. Unfortunately there is a possibility that cereal pollen and wild grass pollen cannot be distinguished in southern India (Thanikaimoni, 1968) so this evidence remains equivocal. This disadvantage does not apply so strongly to north-west India, where Singh *et al.* (1974) found grains of cerealia-type at three sites dating back to 9260 ± 115 B.P. at their earliest in what is now the Rajasthan Desert. These were accompanied by the first significant quantities of carbonised plant fragments which were taken to indicate burning of scrub by man (Singh, 1971). Evidence of more recent cultivation is widespread in India, particularly since about 4000 B.P. (e.g. Vishnu-Mittre, 1966, 1972). There is also good palynological evidence of forest clearance in north India. This has been shown by Sharma and Singh (1972a, b) to have accelerated in one area from c.1200 B.P. Unfortunately there is no clear evidence of forest clearance from south India, although it is almost certain this has occurred. In Sri Lanka much of the grassland is thought to be due to man's influence (Holmes, 1951).

Evidence of early agriculture has also been claimed from northern Thailand, as a result of the excavations at Spirit Cave by Gorman (1972). Although it is not clear which taxa were recovered from which horizon, all the finds appear to come from strata with radiocarbon ages between 7500 and 11 800 B.P. While several of the plants were presumably collected from the wild and brought into the cave, others may have been cultivated. The most likely candidates for this are the bottle gourd (*Lagenaria*), the cucumber (*Cucumis*), the water chestnut (*Trapa*) and the leguminous beans (*Phaseolus, Glycine* or *Vicia*). At present it is impossible to decide whether or not cultivation was occurring and the taxonomic determinations are unconfirmed (D. Yen, personal communication). Another site in northern Thailand, Ban Chieng, has yielded rice husks from an early Bronze Age horizon, 6000–4000 B.P. (Dickenson and Porter, 1972). Again it is difficult to be sure that cultivation was occurring, although it seems likely that it was.

The evidence from island South-East Asia is equally intriguing, for here we have good evidence of early forest clearance, and in some areas agricultural activity, but considerable doubt as to which crops were being grown. In Taiwan, Tsukada (1966, 1967) has shown probable forest clearance back to c.4000 B.P., and suggests possible earlier disturbance of the vegetation, perhaps by shifting cultivation, back to c.10 000 B.P. Evidence for these earlier changes must be regarded as equivocal at present.

Pollen diagrams recently obtained from Sumatra show surprisingly clear indications of forest disturbance. The diagram from Lake Padang (Morley, 1976) at 950 m has already been mentioned (*Figure 5.20*) from the point of view of vegetational change relating to climate. It is, however, equally significant in relation to human activity. Three pollen curves are of particular significance here – those for *Trema, Arenga* and Gramineae. *Trema* is, throughout island South-East Asia, one of the regrowth tree genera. Species of it occur as rare trees in primary forest and must have proliferated particularly since the arrival of man. The *Trema* curve on *Figure 5.20* shows a 'background' presence in the early part of the diagram, but exhibits dramatic increases from an interpolated age of c.5000 B.P. Somewhat later, perhaps from c.4000 B.P., Gramineae pollen grains appear and subsequently increase in abundance. These possibly indicate general deforestation, and the possibility that they include rice pollen cannot be dismissed. *Arenga* pollen first appears at the same level. *Arenga* includes the sugar palm, a wild taxon in Sumatra but one much tended and encouraged by man at the present time. The pollen diagram from the nearby Sikijang Swamp at 1050 m (Flenley *et al.*, 1976) broadly confirms these findings, and indicates their regional, rather than local, nature. The existence of prehistoric man in the area is confirmed by the finds of stone implements nearby (van der Hoop, 1940).

That this early forest clearance was by no means confined to one area of Sumatra is suggested by the findings of Maloney (in Flenley *et al.*, 1976) on the Toba Plateau of North Sumatra at 1400 m. Here small crater swamps provided ideal pollen sites. One of these, Pea Sim-sim, extended back its record to before 18 000 B.P. The upper part of the pollen diagram shows strong evidence of forest clearance, and the macro-stratigraphy shows that a swamp forest has been destroyed (Maloney, 1975). Publication of full details is awaited with interest.

There have been, unfortunately, practically no modern archaeological surveys in Sumatra, with which these findings might be correlated. It is clear, however, from excavations in Sulawesi (Mulvaney and Soejono, 1970; Glover, 1969, 1973) that the Malesian archipelago was by no means a backwater of early civilization.

By far the most striking recent evidence of early agriculture, speaking on a pan-tropical and even a world scale, has come from the New Guinea Highlands. The present large populations of the valleys at c.1500 to c.2500 m there subsist largely on sweet potato (*Ipomoea batatas*) using a variety of rather elaborate

Figure 7.6 Oblique aerial view of Lake Birip, new Guinea, 1900 m. The lake lies in a crater at least 2500 years old. The pollen diagram (below) shows evidence of accelerated horticultural activity beginning about 300 years ago. The trees inside the crater are mostly Casuarina oligodon, planted on old garden sites. In the foreground are numerous mounds used for growth of sweet potato, Ipomoea batatas. (Photography by J. R. Flenley)

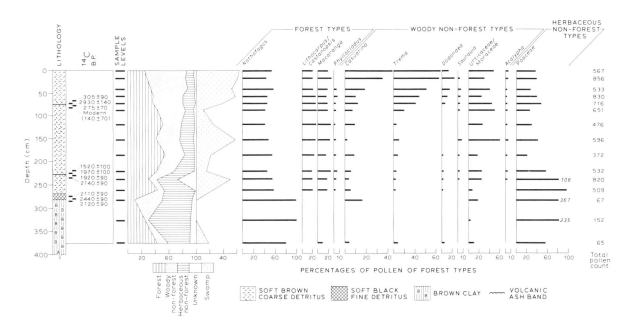

Figure 7.7 Pollen diagram from Lake Birip, New Guinea Highlands (1900 m, 5° 34'S, 143° 50'E). Only selected taxa are shown. Values are percentages of total dry land pollen. (After Flenley, 1967; Walker and Flenley in press)

techniques for its cultivation (e.g. Brookfield and Brown, 1963; Waddell, 1972). An extensive suite of minor crops (Powell, 1976) includes taro (*Colocasia* sp.), yams (*Dioscorea* spp.), bananas (*Musa* spp.) and a variety of recently introduced European vegetables. The chief domestic animals are the pig, the dog and the fowl. Since it was formerly commonly believed (Yén, 1963; Conklin, 1963) that the sweet potato had reached the area from South America in post-Columban times, it used to be thought that the development of elaborate agriculture was fairly late (Watson, 1965a, b).

A pollen diagram from Lake Birip, a crater c.2500 years old, at 1900 m (*Figures 7.6* and *7.7*) showed that disturbance of the vegetation had been in train for probably most or all of the time since deposition began (Flenley, 1967). The early part of the diagram (4.0–2.5 m) is believed to show early stages in regrowth development after the cessation of vulcanicity (*see* Chapter 6). But before forest could be fully established, clearance apparently began, and continued thereafter with increasing intensity until the present day. Un-

fortunately no crop pollens appear, but the behaviour of the *Casuarina* curve is noteworthy. *Casuarina oligodon* occurs naturally along rivers in the area, but is planted everywhere as a source of timber and because of its property, well-known to the New Guineans, of improving soil fertility; actually this is due to nitrogen fixation (Aldrich-Blake, 1932). The pollen diagram suggests this planting has been increasing in popularity for some hundreds of years. The exact date of its beginning is difficult to determine because the upper part of the deposits had been disturbed and radio-carbon dates were correspondingly confused. This disturbance was, however, itself indicative of human activity, for the marginal swamp of the lake is used for cultivation of *Eleocharis sphacelata*, the essential component of the magnificent rustling skirts of New Guinean women. At a higher altitude in the same area there was evidence, at Lake Inim (2550 m), which indicates that forest clearance began only very recently, but that some forest disturbance extends back to an interpolated date of c.1600 B.P. (Walker and Flenley, in press).

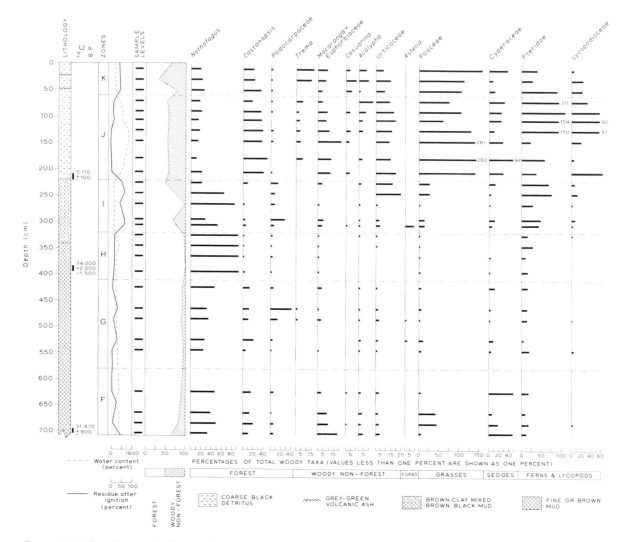

Figure 7.8 Pollen diagram DR29 from Draepi Swamp, New Guinea Highlands, altitude 1885 m. Only selected taxa are shown. Values are percentages of total woody taxa. (After Powell et al., 1975)

Figure 7.9 An excavation at Kuk Swamp, New Guinea, c.1600 m. The upper peat has been removed exposing the grey clay. Cut into the surface of this can be seen two systems of ditches (see page 126). The earlier system consists of a single diffusely-edged ditch running from front to back of the picture. This has been cut across by several of the later sharply-edged ditches, in which wooden digging implements were found (Photograph by J. Golson)

Figure 7.10 Kuk Swamp, New Guinea, c. 1600 m. Two New Guinean men beside the trench which they have dug using wooden digging implements excavated from the peat. Implements of this type have been used in the New Guinea Highlands for at least the last 2000 years. (Photograph by J. Golson)

Then came the first really dramatic evidence, from Draepi Swamp at 1885 m on the northern edges of the Wahgi Valley, the most extensively occupied highland valley in Papua-New Guinea. From here Powell (1970; Powell *et al.*, 1975) produced pollen diagrams spanning the last 30 000 years. One of these (*Figure 7.8*) strongly suggests that an earlier forest-dominated vegetation (Zones *F, G* and *H*) gave way via a transition, Zone *I* (suspected of representing a hiatus in deposition) to an essentially deforested landscape by 5110 ± 100 B.P. (start of Zone *J*). This was indicated by great increases in the percentages of pollen of woody non-forest taxa (including Urticaceae, *Macaranga* and *Trema*) and of Poaceae (Gramineae). This conclusion was confirmed and extended to a higher altitude by the evidence of similar changes dating from c.4500 B.P. at Sirunki (*see* Chapter 5). Neither at Sirunki nor Draepi, however, was there any palynological indication of what crops were being grown, but it seemed unlikely that sweet potato could have been introduced so early, and taro was suggested as a possible staple at the time. This idea received further support when, during the drainage of a swamp at Manton's tea plantation in the Upper Wahgi valley, a series of old drains was discovered in the swamp. These contained well-preserved paddle-shaped wooden implements, presumably used in cutting the peat to make the drains. One of the wooden paddles was dated at c.2300 ± 120 B.P. Since taro is one of the few crops which flourishes in marshy ground, it must be a likely candidate as the crop grown in such situations (Powell *et al.*, 1975). On the other hand, the altitude of Sirunki (2500 m) is more favourable to sweet potato than to taro.

Further exciting evidence has recently emerged from excavations at Kuk Swamp (*Figures 7.9* and *7.10*), also in the Upper Wahgi valley, where five different systems of drainage ditches have been detected at different levels in the deposit (Powell *et al.*, 1975; Golson and Hughes, 1976; Golson, in press; Powell, in press a):

Phase 1, c.9000 B.P.
This consists of ditches, gutters, hollows, pits and stakeholes excavated in the organic deposit. They are thrown into clear relief by the deposition in and above them of a grey clay. The exact purpose behind their construction is unknown but it was probably horticultural. The grey clay deposit is interpreted as resulting from soil erosion on dry land following vegetation destruction by man. Similar (but thinner) inorganic layers in British mires have been given the same interpretation (Simmons *et al.*, 1975).

Phase 2, c.6000 to c.5500 B.P.
This consists of large ditches, up to 2 m wide and deep, and traced for up to 2 km horizontally, with associated complexes of small islands and channels. These features show clearly because they are excavated into the upper surface of the grey clay. It is suggested that dry land crops were grown on the islands and crops such as taro (*Colocasia*) in the channels.

Phase 3, c.4000 to c.2500 B.P.
This is the first true drainage system, consisting of main channels and subsidiary drains. There must have been an elaborate farming of the swamp at this stage.

Phase 4, c.2000 to c.1200 B.P.
This was an even more intensive drainage régime. It is suggested that the main crop grown was taro (*Colocasia*). The abandonment of this system could have been due to the introduction of the sweet potato (*Ipomoea batatas*), which could out-yield taro, but had to be grown on well-drained soils.

Phase 5, c.400 to c.100 B.P.
This consists of extensive and elaborate drains and small field ditches in a gridiron pattern. The latter have been traced on aerial photographs over much of the Upper Wahgi valley floor, and are the same as those used for cultivation of sweet potato on moist soils today.

It is too early to draw firm conclusions about which crops were being grown when and where in this sequence of events. Nor is it easy to explain with certainty the apparent abandonment of successive drainage systems. It is possible that a progressive increase in wetness of the swamp made drainage ever more difficult, and it is also likely that warfare led to the abandonment of some systems. The most recent relinquishment of the swamps could have been due either to the introduction of sweet potato or the introduction of malaria, which may have made the swamplands unhealthy and simultaneously reduced the population (Nelson, 1971).

The most remarkable conclusion is that agriculture was practised in New Guinea by 9000 B.P. This date receives some support, however, from the presence in New Guinea of the pig, an introduced and domesticated species, by c.10 000 B.P. (Bulmer, 1975).

The date of introduction of sweet potato is still obscure, and it is not even certain whether it was pre- or post-Columban. The suggested c.1200 B.P. does not conflict with studies on variation in present-day *Ipomoea* species (Yen, 1973). Also it is conveniently after the arrival of man on the Polynesian islands which are presumed to have been the stepping-stones in the transfer of sweet potato from South America (Howell, 1973). The Polynesians were and are such remarkable navigators (Lewis, 1972) that there seems no difficulty in their crossing the Pacific at an early date.

7.5 CONCLUSIONS

It will be clear from the foregoing that man has indeed been altering the vegetation of equatorial regions for a very long time and in diverse locations. His most striking effect on the vegetation appears to have been forest clearance, and his reason for this clearance has usually been in order to carry out some kind of agriculture. This has been going on since at least 3000 B.P. in Africa, 7000 B.P. in South and Central America, and possibly since 9000 B.P. or earlier in India and New Guinea. We have so far no real evidence as to whether these were independent developments or whether the idea of agriculture diffused from a single centre. In view of the different crops domesticated in each area, independent origin seems likely; on the other hand, the evidence of early agriculture in South-East Asia fits remarkably well with Sauer's (1952) diffusionist ideas.

8
Conclusions, Present Trends and Prospects

8.1 CONCLUSIONS

Any attempt to draw general conclusions about the vegetational history of several separate continental areas is in severe danger of overgeneralisation. Nevertheless, the volume of data available, although not great, is sufficient to demand that the attempt be made. Besides, there are some really rather striking consistencies in the fossil record.

Firstly the indications that equatorial vegetation at any one place was different in the Pleistocene are now all but universal in sequences extending back more than 10 000 years. This may seem an obvious point, but it is worth making, because so much earlier ecology was based on the idea of the unchanging rain forest. The *fact* of change, even in some areas of the lowlands, can now be taken as established.

Secondly we may consider the nature of the former vegetation. In the period c.20 000 to c.14 000 B.P. the vegetation was strikingly different from that at present. In general the mountain vegetation was of types now found at higher altitudes. The lowland vegetation was generally that of areas with a more pronounced dry season.

In the period c.14 000 to c.7000 B.P. vegetation gradually migrated to its modern locations, and probably some new communities developed at the time, particularly in the upper forests on mountains. There were some oscillations of vegetation during this transition period. These could be related to climatic oscillations, but other possibilities cannot yet be excluded. For instance, the new forests formed in previously unforested montane areas were often of few species initially, and may, therefore, have been very subject to pest epidemics. In other areas it is possible that vulcanicity at this period disrupted the usual succession.

Equatorial successions in general have been more variable and less predictable than was formely thought. For example hydroseres have often been interrupted by geological events or human disturbance.

Man-made vegetation has become increasingly abundant in the last few thousand years, and signs of early forest disturbance go back to before 9000 B.P. These early disturbances were temporary and brief, but may have been the result of shifting cultivation. It is even possible than they represent early experiments in agriculture, and it is by no means impossible that agriculture originated in equatorial regions, perhaps at a time when their climate was radically different from at present.

Thirdly, the vegetational history provides strong evidence for Quaternary climatic changes of the most marked kind in equatorial regions. In general, the Late Pleistocene period was cooler and drier than the present day.

The evidence for cooling is the depression of altitudinal vegetation limits in the mountains, summarised in *Figure 3.24* for Africa, *Figure 4.26* for South America, and *Figure 5.23* for New Guinea. This kind of depression has now been traced even in the lowlands, in Panama (p. 72) and Sumatra (*Figure 5.20*). We do not yet have enough information for reliable reconstruction of lapse rates.

The evidence for dessication is from many areas especially the East African Plateau (*Figure 3.20* and *3.21*), Australia (*Figure 5.22*) and South America (*Figures 4.14, 4.20–4.23* and *4.25*), and consists mainly of records of vegetation characteristic of drier climates than occur at the sites today. It is chiefly lowland evidence, but there is some evidence from mountain areas also.

The old 'pluvial' theory must therefore be replaced with a new one in which the glacial periods of temperate areas are reflected in equatorial regions as periods much cooler and more arid than the present.

Fourthly, it may be concluded that rain forest, since it is subject to such changes, cannot be the paragon of stability which it has been held to be. This is not to say that rain forest is necessarily *inherently* unstable, so that if climate continued unchanged for a long time rain forest would be unable to maintain itself. What I am saying is that rain forest in practice *has* changed a great deal in reponse to climate changes at least in some places, and that therefore we cannot assume it is inherently stable.

Fifthly, it follows from this that those explanations of the diversity of rain forest which assume its lack of change over long periods cannot be maintained. The advantage is given to those theories which are indifferent to stability of the forest or actually require isolation of rain forest elements due to climatic change. The possible significance of isolation, leading to speciation in the South American rain forest during the Pleistocene has already been raised, and it seems likely that Africa and perhaps Indo-Malesia experienced similar events of equal significance.

It is, however, difficult to see how this can be the complete explanation of diversity, for many temperate communities also suffered isolation in the Pleistocene. It is a necessary corollary of high diversity that there are many rare species. There simply is not room in the community for many species if they are all to be common – particularly if they are trees, which take up a lot of room. The question 'why is the rain forest so diverse?' may therefore be rephrased 'why are there so many rare species in the rain forest?'. That rarity should be so common leads to the suggestion that there might actually be an advantage in being rare in the rain forest (Flenley, in press). This is so strongly against the usual concept of evolution that it seems impossible at first sight. In fact, however, at least three ways in which rarity can be advantageous have already been suggested. Firstly, there is the idea of pest pressure, introduced by Gillett (1962), and followed up by Janzen (1970). Insect species (and other pests, including fungi) tend to be restricted as to which plant species they will attack; rare species are difficult to find in the forest, and may thus escape serious attack. At the other extreme, monocultures are notorious for susceptibility to insect or fungal attack, and in some of the rare cases of single species dominance in the rain forest, pest epidemics are reported (Anderson, 1961).

The second example is the production of root exudates which inhibit seedling establishment beneath the parent species. This has been demonstrated for the Queensland rain forest tree *Grevillea robusta* by Webb *et al.* (1967a). Clearly, in such a situation, the commoner a tree becomes, the less opportunity it has in the following generation, while the rarer it becomes, the more land is potentially available to it in the next generation.

The third example is that of genetic drift. It has been argued by Fedorov (1966) that many rain forest trees are so rare that they are genetically isolated, or at least belong to very small isolated populations. In these conditions mutations which would be submerged in a larger gene pool may survive, leading to that genetic variability which is advantageous for survival. If this mechanism were continued it would lead to eventual reproductive isolation and hence to speciation, producing yet more rare species. The existence of this mechanism has been disputed (Ashton, 1969), but it would certainly be more likely to occur in rare species than common ones.

The generalisation that rarity is an advantage can only be an explanation of diversity if it is more effective in the wet tropics than in most temperate or arid regions. It will be noted that at least the first two of the examples quoted apply particularly to seedling survival. Perhaps seedling survival in temperate and arid regions is controlled largely by 'across the board' climatic effects such as temperature and soil water content, which would exhibit no selective effect in favour of rare species.

Sixthly, the evidence of abundant change in the equatorial vegetation is one more nail in the coffin of the climax theory. Clearly we cannot assume that a forest which has existed for only a few tree generations is in a state of stable equilibrium with its environment. On the contrary, the evidence suggests that the forest is still changing, and that Quaternary climatic variations are sufficient to prevent any climax phase being reached. The climax may exist as a theoretical ideal but in the equatorial regions it probably does not exist in practice at the present day.

Seventhly and lastly, we may draw limited conclusions as to whether equatorial vegetation behaves in a holistic manner (Clements, 1916; Tansley, 1920) or whether it is essentially individualistic (Gleason, 1926). It has been pointed out by Livingstone (1975) for Africa and by D. Walker (personal communication) for New Guinea that the pollen record suggests a good deal of individualistic behaviour. This is particularly shown by the rate of immigration of taxa to an area after the last Pleistocene period of climatic adversity. There is certainly no simple upward movement of forest limits; it would be nearer the truth to say that one observes the gradual synthesis of the montane forest that we know today. Perhaps the most striking example of this is the late arrival of *Podocarpus* spp. in many African montane forests. In New Guinea the forest has still not settled down to uniform composition, and every peak has its own vegetational peculiarities (Flenley, 1967). Indeed, it seems likely from a consideration of Pleistocene climatic trends established by other means, that the upper montane forest taxa are at present enjoying no more than a brief excursion to high altitudes before returning to their normal Quaternary home below 2500 m (Walker and Flenley, in press).

8.2 PRESENT TRENDS

The historian usually closes his account a respectful distance before the present, and the sensible vegetational historian should probably do the same. Perhaps two pleas may be entered for doing otherwise. Firstly, present trends in vegetational change are so striking that they cannot lightly be ignored; secondly, plants are not quite as capricious in their behaviour as human beings.

Very short-term climatic fluctuations are having some influence on sensitive vegetation in equatorial regions. One of the most noticeable examples of this is the influence of occasional periods of severe frost in the Highlands of New Guinea. One such occurred about 1940, and a further one was recorded in detail in 1972 (Brown and Powell, 1974). As well as causing extensive damage to sweet potato gardens, producing a famine, the frost severely affected native trees in areas subject to cold air drainage. Frost is a widespread phenomenon in the montane tropics (van Steenis, 1968).

By far the greatest changes in equatorial vegetation at present are due to man. So many claims have been made about this in recent years that it is difficult to separate hard fact from exaggerated evidence. We shall therefore consider a few examples of different types of acceptable evidence. In some cases it is possible to reconstruct former vegetation patterns with reasonable accuracy from historical evidence. There are neat cases of this from Trinidad (Fonaroff, 1974; *Figure 8.1*) and Barbados (Watts, 1970; *Figure 8.2*). It is clear in both examples that while the major clearances occurred a long time ago, other changes, particularly connected with agriculture, are much more recent. It is extraordinarily difficult to obtain accurate estimates of rates of forest clearance, but a good attempt has been made by Marshall (1973) for West Malaysia (*Figure 8.3*). It is evident that there has been a striking acceleration in the rate of clearance in recent years, and that there is a close parallel with the rate of population increase. Similar evidence over a longer period is available from Sri Lanka (Crusz, 1973; *Figure 8.4*). Unfortunately it is not clear from the graph how much of the area is still forested, nor whether creation of new paddy land (for rice growing) involves removal of forest. A high proportion of the land is already covered with man-made grassland (Holmes, 1951; Perera, 1968). The relationship between population and cleared land in *Figure 8.3* and between population and paddy land in *Figure 8.4* suggests that these are closely related. In view of the need of a growing population for more agricultural land there seems no reason why this should not be a causal relationship. It should not be imagined that

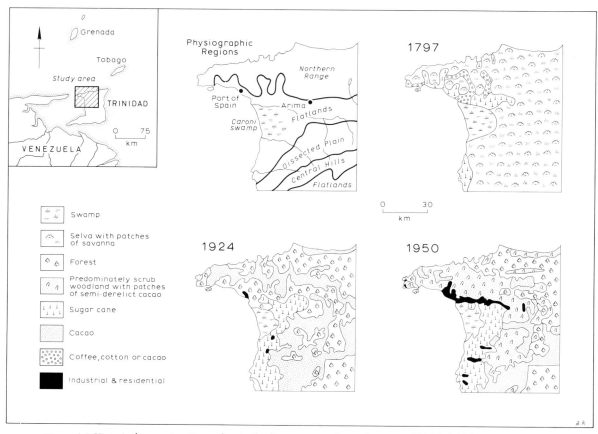

Figure 8.1 Historical reconstructions of major land use patterns in north-western Trinidad. (After Fonaroff, 1974)

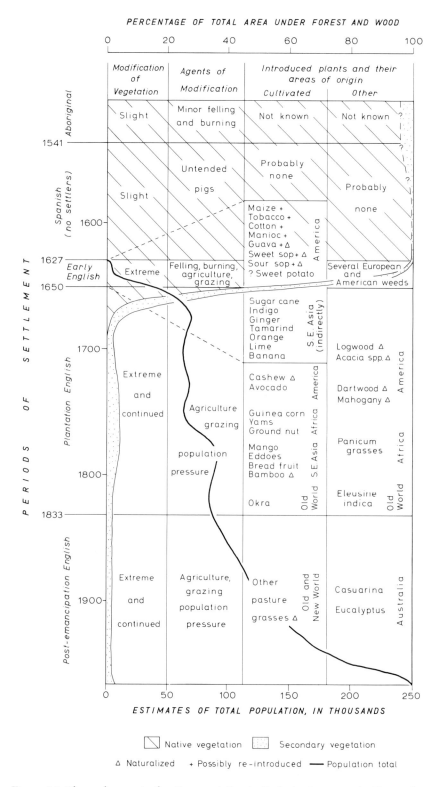

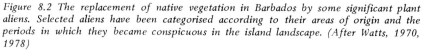

Figure 8.2 The replacement of native vegetation in Barbados by some significant plant aliens. Selected aliens have been categorised according to their areas of origin and the periods in which they became conspicuous in the island landscape. (After Watts, 1970, 1978)

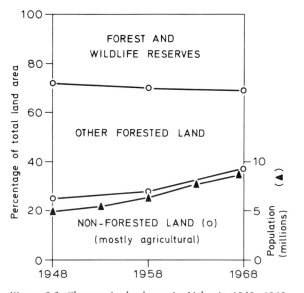

Figure 8.3 Changes in land use in Malaysia 1948–1968. (Data from Marshall, 1973)

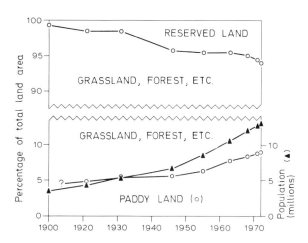

Figure 8.4 Changes in land use in Sri Lanka, 1900–1972. (Data from Crusz, 1973)

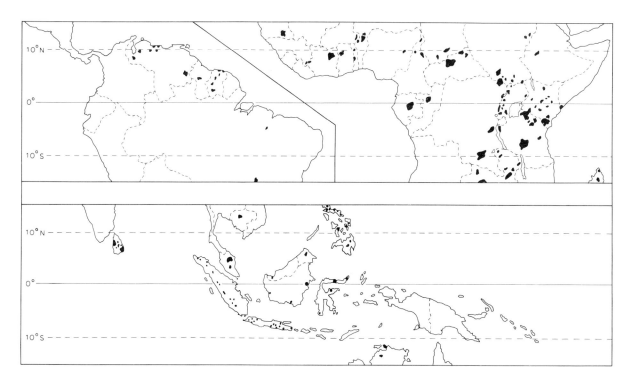

Figure 8.5 Nature reserves and equivalent areas in equatorial regions. Miller cylindrical projection. (Data from UN list of national parks and equivalent reserves, 1971)

the relationship would be a simple one in all countries, however.

The most remarkable case of vegetational change in recent decades was perhaps that during the Vietnam War. Although this example is outside the strictly equatorial regions it is of such magnitude that it cannot be ignored. The facts have been concisely presented by Westing (1972). Between 1961 and 1971 the USA expended over 49 M kilograms of herbicides (measured as active ingredients) on 2M hectares of forest, and over 3M kilograms on 300,000 hectares of croplands. The chief herbicides used on the forest were 2,4–D and 2,4,5–T, with a smaller amount of Picloram. On cropland dimethylarsinic acid was used. 15% of the forested land of South Vietnam was sprayed once, and an additional 4% multiple-sprayed. 8% of the cropland was sprayed. The spraying was extremely effective in defoliating the trees and frequently in killing them. Westing estimated that recovery time after one spraying of forest is more than a decade, and after multiple-spraying is several decades. These are presumably estimates for recovery to a forest canopy. It is inevitable, however, that some species will be more sensitive than others, as any gardener who has treated a lawn knows. To recover the full floristic diversity of the forest may therefore be a very much longer process.

Detailed records of vegetational change are best obtained by means of permanent quadrats, and a number of these are now being maintained in equatorial regions. In the Galapagos Islands changes could be observed even in the short period 1966–1973; most striking was the rapid recovery of vegetation on the Isla Santa Fe since the elimination of goats in 1971 (Hamann, 1975).

Another method of assessing short-term changes in vegetation is by the use of aerial photographs and satellite photographs. The latter are adequate for distinguishing forest from non-forest, while good aerial photographs allow quite elaborate distinctions of vegetation types to be made. These methods have been applied in South Sumatra (Sobur *et al.*, 1975). The area concerned was photographed from the air in 1969 and 1975, and from space since 1973 by the Earth Resources Technology Satellite. It is a resettlement area, and in the 6 years 1969–1975 about 35% of the primary swamp forest disappeared to agriculture or other uses. The continuation of satellite photography should allow us to make accurate estimates of rates of forest destruction within the next few years for any desired area.

The acceleration of vegetational change has given rise to increasing efforts to set aside reserves. The trend can be seen in *Figures 8.3 and 8.4* and is well known in other areas. The distribution in 1971 of reserves in equatorial regions is summarised in *Figure 8.5*. A number of reserves are in the large 'national park' category, but many are very small. This is unfortunate for two reasons. Firstly, the marginal areas of any reserve are subject to influence from outside. In forest this edge-effect often amounts to no more than the entry of light into the otherwise shady interior of the forest, but it is sufficient to influence the vegetation. Other edge-effects include danger of fire, land drainage, and incursion of domestic animals. In small reserves edge-effect influences a significant proportion – or even all – of the reserve. Secondly, the reserve is essentially an island in a sea of different vegetation, and is therefore subject to the biogeographical laws governing islands. The diversity of an island flora is a balance between immigration of new species and extinction of those already there (Mac Arthur and Wilson, 1967). Small islands can support only an impoverished flora, while larger ones can support a much richer one, because the chances of extinction are reduced in the larger area available. Large reserves have the same advantage (Hooper, 1971; Diamond, 1975).

8.3   PROSPECTS

Only fools and seers prophesy, and I make no bid to be either. We may, however, look briefly at what *could* happen in the unlikely event of present trends continuing.

The possible effects of long term climatic fluctuation on vegetation in the future can be predicted only if one accepts the astronomical theory of climatic change (Hays *et al.*, 1976). This theory predicts that we are already part of the way towards another temperate zone glaciation and, in tropical regions, another dessication.

The possible future effects of vegetational change brought about by man can be predicted with a little more confidence. A glance at *Figure 8.3*, for example, will show that even without further acceleration of deforestation (which has probably already occurred), there would be nothing but non-forest and reserve left in West Malaysia by the year 2000. In such a situation it seems likely that economic pressure would force some reserves to be made available for clearance, but it should be pointed out that many of the reserves in Malaysia are on mountainous terrain; clearance of these would probably be economically counter-productive because of increased runoff causing flooding in the lowlands (Douglas, 1971). Malaysia is not atypical. Some equatorial countries have more forested land, but others, such as the Philippines, probably have less, and the immediate, if temporary, economic rewards of logging are high almost everywhere.

Tropical forest silviculture formerly used a polycyclic system which involved removal of selected canopy trees at fairly frequent intervals. Since the 1950s economic changes have led to the adoption of a monocyclic system in which all saleable trees are removed at a single operation (Whitmore, 1975). Planting of young trees is sometimes resorted to after a monocyclic felling. Irregular ecological events, such as cyclones and insect attack, can easily play havoc with the monocyclic system, as they do to forestry plans in temperate regions (Raup, 1964). Even if silviculture succeeds in producing a 'sustained yield' in quantity of timber, this is very different from a sustained economic yield, because of fluctuations in price. Economic and social needs therefore often conflict with the silvicultural ideal.

It is now all but inevitable that much more deforestation will take place in equatorial regions by the end of the century, and it has been suggested that this could have profound ecological effects. For instance it has been claimed that destruction of the major rain forest area, particularly in the Amazon basin, could reduce the conversion of carbon dioxide to oxygen by photosynthesis and thus reduce the oxygen content of the atmosphere. This seems very unlikely, for all the deforestation that has already occurred has had no measurable effect on the value of 21% oxygen. A more likely effect is that on the climate. A computer simulation has suggested the following chain of events:

'deforestation → increased surface albedo → reduced surface absorption of solar energy → surface cooling → reduced evaporation and sensible heat flux from the surface → reduced convective activity and rainfall → reduced release of latent heat, weakened Hadley circulation, and cooling in the middle and upper tropical troposphere → increased tropical lapse rates → increased precipitation in the latitude bands 5-25°N and 5-25°S and a decrease in the Equator—pole temperature gradient → reduced meridional transport of heat and moisture out of equatorial regions → global cooling and a decrease in precipitation between 45° and 85°N and at 40° and 60°S' (Potter *et al.*, 1975).

It is believed by some that the destruction of rain forest is an irreversible process. Under present conditions where firing and agriculture prevent regeneration of the forest this is probably true. The vegetational history shows that although rain forest can migrate and cover whole landscapes, it may take several hundred years to achieve full diversity. Forest clearance should not, therefore, be undertaken lightly.

It may be argued that the abundance of rain forest in recent millenia, though perhaps normal on a 50 M year scale, is rather abnormal on a 2 M year scale, because of the generally adverse climates of the Pleistocene. It is likely, however, that deforestation in a moist warm climate such as we have at present can have harmful effects — especially on the soil — which were not produced by lack of forest during the Pleistocene times of dry, cool climate. If present rates of felling continue, the amount of forest left will soon be much less than that during the most adverse phases of the Pleistocene. It is clear that if matters are allowed to continue at the present rate then very serious problems will arise. Man has often been urged to learn the lessons of history; he must also learn the lessons of vegetational history.

Appendix 1

NOTES ON COMMON POLLEN TYPES OF THE
EQUATORIAL EAST AFRICAN QUATERNARY

The information here is mostly from Hamilton (1972) and Livingstone (1967).

Acalypha (Euphorbiaceae). Common in moist and dry lowland forest, and the pollen exhibits very high relative export. Abundance of this pollen type therefore indicates forests in the lowlands, but not necessarily near the sampling site.

Afrocrania (Cornaceae). This pollen type is shown by Hamilton to be of low relative export, and is therefore a good indicator of montane forest growing nearby.

Alchemilla (Rosaceae). The pollen is of poor relative export and therefore its presence indicates proximity of *Alchemilla* spp. These shrubs are only really abundant in the Afro-alpine Belt, but some species do occur in swamps of much lower altitudes.

Anthospermum (Rubiaceae). It is characteristic of the drier mountains and may be taken as an indicator of dryness. The pollen is of moderate relative export.

Artemisia (Compositae). *A. afra* of the Afro-alpine and Ericaceous Belts is probably responsible. It is characteristic of the drier mountains and may be taken as an indicator of dryness. The pollen is of moderate relative export.

Celtis-type (Ulmaceae). *Celtis* spp. are common in moist and dry lowland forests and the pollen exhibits very high relative export. Abundance of this pollen type therefore indicates forest in the lowlands, but not necessarily near the sample site.

Chenopodiaceae. An indicator of human disturbance.

Dendrosenecio-type (Compositae). Species of tree *Senecio* are common in the Afro-alpine Belt but also occur down to 2600 m and thus cannot be used as a clear indicator of Afro-alpine conditions. Also, the pollen shows moderate relative export.

Dodonaea viscosa (Sapindaceae). A shrub indicating human disturbance, when abundant.

Ericaceae. Most of the pollen probably comes from *Phillippia* and *Erica* spp. in the Ericaceous Belt, but species of *Vaccinium* and *Blaeria* may also contribute. The pollen is of moderate relative export.

Gramineae. So many species may contribute to the Gramineae curve that it is difficult to interpret. The pollen is of high relative export so it is also difficult to be certain of the position of the source in relation to the sampling site. High values for Gramineae are consistent with Afro-alpine vegetation, with disturbed vegetation at any altitude, with savannah, or with swamp vegetation.

Hagenia (Rosaceae). *H. abyssinica* pollen is of moderate relative export, and the parent tree, although altitudinally restricted to the uppermost Montane Belt and the lower Ericaceous Belt is found on both moist and dry mountains. It is, therefore, of limited value as an indicator of ecological conditions.

Hypericum (Guttiferae). The pollen is of low relative export, and derives from several species of *Hypericum*, occurring from below 2000 m to over 4000 m altitude, generally in drier situations.

Lobelia (Campanulaceae). There are herbaceous lobelias as well as the tree species, and they grow not just in the Afro-alpine Belt and upper Ericaceous Belt but at a wide variety of altitudes. They are, therefore, of little value as altitudinal indicators, even though the pollen is of low relative export.

Macaranga (Euphorbiaceae). *M. kilimandscharica*, the likely pollen source, is characteristic of wetter forests in Kenya, and also occurs in secondary vegetation. The pollen shows high relative export.

Myrica (Myricaceae). Two species may be involved: *M. kandtiana*, a swamp forest tree, and *M. salicifolia*, which is a common tree on dry sites in the Montane Forest. The pollen exhibits moderate relative export, and the interpretation of which species was responsible must depend on other factors, such as whether there is independent evidence of the existence of swamp.

Myrsinaceae. Some workers have distinguished *Rapanea* pollen, but others have combined all Myrsinaceae together. *Rapanea* is particularly abundant in the upper Bamboo Zone and forms a distinct belt with *Hagenia* in some mountains (e.g. Ruwenzori). The pollen varies widely in the degree of export exhibited.

Olea (Oleaceae). There are several species of olive trees in both lowland and dry montane forests. The pollen exhibits high relative export.

Podocarpus (Podocarpaceae). There are several species of the genus in East Africa; but the most important are *P. gracilior* and *P. milanjianus*. Both are indicators of the dry type of montane forests. The pollen exhibits very high relative export and is not, therefore, to be taken as evidence of the close proximity of the source forest.

Stoebe kilimandscharica (Compositae). This shrub is fairly characteristic of the Ericaceous Belt and just above and below it, especially in dry localities (Hedberg, 1957). The pollen exhibits moderate relative export, but has some indicator value.

Urticaceae. Urticaceae species are characteristic of moist upland forests, and the pollen is in the high relative export class. Abundance of this type suggests the existence of moist upland forests, but not necessarily near the sampling site. This characteristic is particularly useful for identifying the Bamboo Zone, which is otherwise a poor pollen producer.

Appendix 2

NOTES ON COMMON POLLEN TYPES AND OTHER
MICROFOSSILS OF THE EQUATORIAL LATIN
AMERICAN QUATERNARY

The information here is mostly from Muller (1959), van der Hammen and Gonzalez (1960), and van der Hammen (1974).

Acaena/Polylepis (Rosaceae). *Acaena* spp. (herbs and dwarf shrubs) are locally common in the Paramo. *Polylepis* spp. are trees of the Andean forest, occurring in isolated patches up to 4000 m and also as shrubs of the Paramo (Schimper, 1903). The pollen is distinguishable under the electron microscope, but not with certainty under the light microscope (van Geel and van der Hammen, 1973). The pollen is only abundant in surface samples at altitudes above 3800 m, and is therefore used as a major indicator in pollen diagrams.

Acrostichum (Pteridaceae). A fern found locally in *Avicennia* forest. The spores probably easily lose their faint ornamentation, and are then included under 'Trilete psilate (large)'.

Alnus (Betulaceae). *A. jorullensis* is a common tree, dominant in wet areas, in the Andean forest. Pollen production is high, and the pollen is wind-dispersed and of high relative export. The habitat of the tree, alongside rivers, also promotes its transport by rivers to the lowlands.

Avicennia (Avicenniaceae). *A. nitida* forms the main component of the back of the mangrove swamp. The pollen distribution is closely related to the source area, but some transport by water occurs.

Borreria (Rubiaceae). Herbs of the forest at several altitudes.

Botryococcus. An alga. The precise ecological controls are not well established.

Byrsonima (Malpighiaceae). A genus of trees including small trees of the savannahs. We have no information on its pollen dispersal properties.

Caryophyllaceae. Herbs of the Paramo and the upper forest.

Chenopodiaceae and Amaranthaceae. These include herbs of the Paramo, of deforested land, and of xerophytic vegetation.

Coelastrum. An alga whose distribution strongly suggests a dependence on relatively high temperature.

Compositae (Liguliflorae). Many genera at various altitudes, but the pollen type only rarely occurs.

Compositae (Tubuliflorae). Many genera at various altitudes including *Espeletia*, which is abundant in the Paramo.

Curatella (Dilleniaceae). A genus of trees including small trees of the savannahs. Although it is very frequent in the Cerrados, its pollen was not found in the pollen rain (Salgado-Labouriau, 1973).

Cyperaceae. Very frequent herbs of the Paramo and marshes. There are also species of the Andean forest, and in the savannahs.

Dodonaea (Sapindaceae). A shrub of cleared land, eroded soils and xerophytic vegetation.

Drimys (Winteraceae). *D. granatensis* L.f. is a relatively common tree, particularly in the higher zone of the Andean forest. We have little information on the pollen dispersal, but some relative export occurs.

Ericaceae. These occur as trees, shrubs and dwarf shrubs in the Andean forest, Sub-paramo and Paramo.

Gentiana (Gentianaceae). Herbs of the Paramo.

Geranium (Geraniaceae). Herbs of the Paramo; occasionally at lower latitudes.

Gramineae. Although Gramineae are abundant in the Paramo, they also occur in deforested land at lower altitudes and as the bamboo genus *Chusquea* centred on the Sub-paramo. The pollen is abundantly produced, wind-transported and of medium relative export.

Hedyosmum (Chloranthaceae). *H. bonplandianum* H.B.K. is a woody climber of the Quercetum, and the genus occurs also as trees in the Weinmannietum and in other forests at all altitudes. The pollen production is heavy and the plant is anemophilous. The pollen is of high relative export.

Hypericum (Guttiferae). Shrubs, particularly of the Sub-paramo. The pollen exhibits some relative export.

Ilex (Aquifoliaceae). A genus of trees and shrubs of the Andean forest. Some species may occur above the forest limit. The pollen appears to be of medium/low relative export.

Isoetes (Isoetaceae). Plants of lakes, marshes and bogs in the high Paramo.

Jamesonia (Gymnogrammaceae). A fern of the high Paramo.

Jussiaea (= *Ludwigia*) (Onagraceae). Marsh plants at various altitudes.

Malpighiaceae (excluding *Byrsonima*). Trees of many forest types. We have no information on their pollen dispersal properties.

Malvaceae. The pollen found in the high parts of the Andes is believed to be that of *Malvastrum acaule*, which occurs 3800–4200 m, and the pollen is, therefore, a useful indicator.

Mauritia (Palmae). This is the genus of palms dominant in the 'Morichales' swamps of the Orinoco delta. The pollen appears to be of low relative export.

Miconia and Melastomataceae. Morphologically the pollen of *Miconia* is not easily distinguishable from other Melastomataceae, although in Colombia van der Hammen and Gonzalez (1960) were able to separate them by measurement. There are many species of *Miconia* and other Melastomataceae in the forest at all altitudes, and even considerably above the forest limit. The pollen appears to be of medium/low relative export.

Myrica (Myricaceae). There are two important species, *M. parviflora* and *M. pubescens*, common trees or shrubs in the Andean forest. The pollen exhibits moderate relative export.

Myriophyllum (Haloragidaceae). An aquatic found mainly in the Paramo, and therefore a good indicator of cold conditions.

Myrtaceae and *Eugenia*. *Eugenia* and other genera of Myrtaceae are trees of the Andean forest, and also occur at lower altitudes. The pollen exhibits some relative export.

Palmae. Palms are of general occurrence in the lowland forest and *Selva subandina*, and become very abundant in palm swamps of the Orinoco delta.

Plantago (Plantaginaceae). Locally very abundant in Paramo; also occurs in cleared areas at lower altitudes.

Podocarpus (Podocarpaceae). Several species of *Podocarpus* trees are found in the mountain forests. The pollen appears to be of moderate relative export.

Polygonum (Polygonaceae). Herbs of the Paramo, upper forest and lower altitudes.

Quercus (Fagaceae). This is the dominant tree genus of the Andean forest in some areas. The pollen shows high relative export.

Ranunculus (Ranunculaceae). Herbs of the Paramo.

Rapanea (Myrsinaceae). Several species of trees or shrubs, which are frequent in the Andean forest. Pollen of medium/high relative export.

Rhizophora (Rhizophoraceae). *R. mangle* is the main outer mangrove species forming the advancing front of the swamp. Other species occur further back in the swamp and in other places. It is anemophilous and a heavy producer of small grains. The pollen exhibits very high relative export, probably by aerial and water transport.

Rumex (Polygonaceae). Herbs of the Paramo.

Symphonia (Guttiferae). *S. globulifera* is common in the lowlands as a riverside and swamp tree. In Trinidad the tree also occurs in evergreen seasonal forest, *Mora* forest, and lower montane and montane forest.

Terminalia-type (Combretaceae). The pollen type occurs in *Buceras, Conocarpus* and *Terminalia* (all Combretaceae); *Terminalia* is by far the commonest of these, with several species of tree in almost all types of forest. The pollen is capable of some relative export.

Thalictrum (Ranunculaceae). A climbing herb of the Andean forest.

Urticaceae. Common shrubs and herbs of the forest at all altitudes. The pollen exhibits medium relative export.

Valeriana (Valerianaceae). Herbs of the Paramo; there is one shrubby species, *V. arborea*, but it has distinct pollen. The pollen is of low relative export.

Weinmannia (Cunoniaceae). The dominant genus of trees in most types of Andean forest, one of the commonest species being *W. tomentosa*. The pollen appears to be of moderate relative export.

Appendix 3

NOTES ON COMMON POLLEN AND SPORE TYPES OF
THE EQUATORIAL INDO—MALESIAN QUATERNARY

The information here is mostly from Powell (1970), Hope (1973, 1976a), Flenley (1967, 1973), Morley (1976), Maloney (in preparation), and Muller (personal communication).

Acalypha (Euphorbiaceae). This genus includes shrubs of secondary vegetation, and the pollen is therefore taken as an indication of forest destruction. The pollen exhibits moderate relative export.

Arenga spp. (Palmae) are palms of the lowland forest, but *A. pinnata* (the sugar palm) is widely cultivated.

Ascarina (Chloranthaceae). *A. philippinensis* is an occasional tree of lower montane forest throughout Malesia. The pollen is of low relative export.

Astelia (Liliaceae). *A. papuana* is the only species of the genus found in Indo—Malesia. It is an element of tropicalpine herbfields and bogs above 3050 m in New Guinea. It is never epiphytic, unlike some New Zealand *Astelia* spp. The pollen may possibly be transported some way uphill, but extremely rarely downwards. It is therefore a good indicator of tropicalpine herbfield or bog, or at least of 'sub-alpine' vegetation.

Casuarina (Casuarinaceae). A genus of trees which are occasionally found in rain forest but may become abundant or dominant in areas with a pronounced dry season, for example North Australia and East Java. They may be seral, e.g. after fire in East Java (van Steenis and Schippers-Lammertse, 1965) and after volcanism in New Guinea (Taylor, 1957). The species are particularly found along water courses in many areas. In addition *Casuarina oligodon* is widely planted on old garden sites in the New Guinea Highlands, and thus becomes a good indicator of human activity there. The pollen (Kershaw, 1970b) is wind-dispersed and of high relative export. It has probably even been carried by wind from Australia to New Zealand (Moar, 1969).

Celtis (Ulmaceae). *Celtis* spp. are locally common trees in lowland rain forest, but do not extend much above 1000 m. The pollen, however, is carried far uphill.

Coprosma (Rubiaceae). Several species of shrub belonging to this genus occur in New Guinea, and the genus extends westwards to Mt Kinabalu and East Java. They are particularly abundant in the sub-alpine zone.

Cyatheaceae. The main family of tree-ferns in Indo—Malesia. Many species are present in the forest and a few are abundant in regrowth over a wide range of altitudes. They are often particularly prevalent in sub-alpine vegetation. The spores exhibit moderate relative export.

Dacrycarpus (*Podocarpus* sect. *Dacrycarpus*) (Podocarpaceae). The 3-vesiculate pollen grains of this genus are quite distinct from those of *Podocarpus* s.s. The genus occurs in the uppermost forests, descending to lower altitudes in swamp forest. The pollen reaches peak values above the forest limit on Mt Kinabalu, and is therefore presumably of high relative export. The altitudinal range in West Malesia is ±1000—

2500 m, but it extends to well above 3000 m in New Guinea.

Dacrydium (Podocarpaceae). Trees of this coniferous genus occur in montane forests in West Malesia and also in swamp forests in New Guinea. On Kinabalu they are dominant about 2400 m. The pollen is of moderate/low relative export.

Dodonaea (Sapindaceae). *D. viscosa* is a small regrowth tree in New Guinea and a shrub of sclerophyll forest in Queensland. The pollen exhibits moderate relative export.

Drapetes (Thymelaeaceae). *D. ericoides* is the only species in Indo–Malesia, occuring on New Guinea mountains almost entirely above 3050 m, and on Mt Kinabalu, Borneo, above c.3000 m. The pollen is not common, but occurs only in tropicalpine spectra, so it appears to be a good indicator of tropicalpine vegetation.

Elaeocarpus (Elaeocarpaceae). *Elaeocarpus* spp. are abundant trees in the forest at a wide range of altitudes, and occur in swamp forest in Sumatra.

Engelhardia (= *Engelhardtia*) (Juglandaceae). A genus of trees of the mountain forest, and to some extent of the lowland forest also. *E. spicata* also occurs in mountain swamps in Sumatra.

Ericaceae. So many taxa are represented (*Rhododendron, Vaccinium, Dimorphanthera* etc.) that it is difficult to assign much significance to the pollen.

Gentiana (Gentianaceae). Although common in tropicalpine communities, *Gentiana* spp. also occur at a variety of altitudes down to 1500 m or below in New Guinea, Java etc. and are therefore not very good tropicalpine indicators. The pollen appears to be of low relative export.

Gramineae (Poaceae). In Indo-Malesia, Gramineae may be abundant in swamps (e.g. *Phragmites karka*) as well as above the forest limit and at all altitudes on dry land which is not forested (e.g. *Imperata cylindrica*). There are also bamboos in the forest. Pollen of rice (*Oryza sativa*) and other cultivars, is unfortunately not distinct from that of wild grasses (Maloney, in preparation). The status of Gramineae in the region has been reviewed by Whyte (1972).

Haloragaceae (=Haloragidaceae). In New Guinea this pollen type probably represents *Haloragis micrantha*; in Sumatra it may include *Laurembergia coccinea* also. Both are sub-aquatic herbs; in Java neither occurs below 1600 m (van Steenis, 1972). *L. hirsuta* occurs down to 1300 m in Sumatra (Morley, 1976).

Ilex (Aquifoliaceae). Trees of rain forest or swamp forest. *I. cymosa* is abundant in swamp forest in Sumatra.

Lithocarpus comp. (= *Lithocarpus/Castanopsis*) (Fagaceae). These are the oaks of the 'oak' forest of New Guinea, and they are also common in lower montane forests (c.1000–2000 m) throughout Malesia. Unfortunately the pollen of the two genera cannot be distinguished (Muller, 1965). *Lithocarpus* is believed to be insect pollinated (Soepadmo, 1972) but *Castanopsis* is wind pollinated, and the pollen type is of fairly high relative export.

Macaranga (Euphorbiaceae). There are many species of this tree genus, chiefly in secondary vegetation, but some in primary forest. The pollen displays high relative export.

Melastomataceae. These are trees or shrubs in forest, or shrubs in regrowth and swamp. It seems likely that only the regrowth and swamp shrubs have a chance to contribute pollen to the pollen rain.

Myrica (Myricaceae). The pollen is similar to *Casuarina* but distinguished by its smaller size, absence of crevassing (*sensu* Kershaw, 1970b) and more pronounced vestibulum (Morley, 1976). There are two species in West Malesia, both trees. *M. esculenta* is a mainly lowland species (0–1800 m) while *M. javanica* is chiefly upland (700–3300 m), and may dominate the montane forest. It is also a pioneer species in vulcanoseres.

Myrsine (Myrsinaceae). *M. avenis* and *M. affinis* are common treelets in swamp forest in Sumatra; *M. avenis* is also reported at higher altitudes in 'elfin and mossy forest' in Java (van Steenis, 1972).

Myrtaceae. This family includes many trees of the rain forest (especially *Eugenia/Syzygium* spp.) at all altitudes. It also includes *Eucalyptus*, the chief Australian sclerophyll genus. The pollen exhibits moderate relative export, and the pollen of *Eucalyptus* is usually distinguishable from that of *Eugenia/Syzygium*.

Nothofagus id. *brassii* sim. (Fagaceae). There are several *Nothofagus* species in the New Guinea Highlands, almost always between 1000 and 3100 m (Hynes, 1974). They are the dominant species of 'beech forest'. The pollen is highly distinctive (Cranwell, 1963) and is wind dispersed. It is capable of exceptionally high relative export, and high values for it in spectra from above the forest limit are quite usual.

Oreomyrrhis (Umbelliferae). In New Guinea *Oreomyrrhis* spp. are only recorded above 3050 m where they contribute to tropicalpine herbfield. They therefore make good altitudinal indicators. The pollen appears to be of low relative export.

Pandanus (Pandanaceae). A genus of monocotyledonous trees or treelets which occur in rain forests and in swamps over a wide range of altitudes. We have no real information about the distributive properties of the pollen.

Phyllocladus (Phyllocladaceae). *P. hypophyllus*, the only species in the area, is a tree of mountain rain forest in New Guinea and Borneo. On Mt Kinabalu it is dominant around 2700 m. It can also contribute to swamp forest. The pollen exhibits very high relative export, and dominates spectra from above the forest limit on Kinabalu.

Pinus (Pinaceae). *P. kesiya* occurs in the Philippines and *P. merkusii* in the Philippines and Sumatra, as far south as the Kerinci Valley (2°S). Pollen of *Pinus* occurs in surface samples from Mt Kinabalu (det. J. Muller), and is presumably wind-blown from the Philippines.

Plantago (Plantaginaceae). *P. major* is abundant in Java and Sumatra in open habitats from 0–3300 m, but is absent from New Guinea, where its place as a weed at low and middle altitudes is taken by *P. varia*. The pollen of both of these is coarsely verrucate. The New Guinea high mountains support four further species (van Royen, 1963) that are characteristic of tropic-alpine vegetation and have distinctive pollen.

Podocarpus (Podocarpaceae). Several species of these trees inhabit the rain forest, and there is a tendency for them to become more abundant at middle altitudes (c.1500 m) or higher. The pollen is of moderate relative export.

Potentilla (Rosaceae). Several *Potentilla* species contribute to tropicalpine vegetation throughout Malesia, but they also occur at lower altitudes (down to 2150 m in New Guinea) and they are therefore not good altitudinal indicators. The pollen is of low relative export.

Quercus (Fagaceae). One of the genera of trees in the lower mountain forest of West Malesia. *Quercus* is anemophilous and the pollen is carried to high altitudes, exhibiting moderate relative export.

Quintinia (Saxifragaceae). This is a montane tree genus, rather characteristic of the uppermost 'mixed' forest in New Guinea. The pollen is highly distinctive and is possibly of moderate relative export.

Ranunculus (Ranunculaceae). Although common in tropicalpine communities, *Ranunculus* spp. also occur at a variety of altitudes down to 1200 m in New Guinea and elsewhere, and are therefore not good altitudinal indicators. The pollen appears to be of low relative export, and high values are obtained from tundra and herbfields.

Rapanea (Myrsinaceae). *R. vaccinioides* is an abundant species in upper montane and sub-alpine forests in New Guinea (Wade and McVean, 1969). Many other species occur in forest at lower altitudes, but the pollen from these does not seem to pass out of the forest. Pollen from the sub-alpine forests is carried well uphill on Mt Wilhelm.

Saurauia (Saurauiaceae). Small trees which occur in primary forest but are far more abundant in re-growth. The pollen is probably an indication of disturbance of the forest.

Styphelia (Epacridaceae). *S. sauveolens* is a dwarf shrub of the tropicalpine zone in New Guinea, Kinabalu etc., but also occurs at lower altitudes (down to 1850 m in New Guinea). Some other Epacridaceae may contribute to the results for this taxon.

Symingtonia (Hamamelidaceae). *S. populnea* is a characteristic tree of forest between 1000 and 3000 m (van Steenis, 1958). In the Kerinci Valley, Sumatra, it was confined to above 1800 m (Morley, 1976).

Trema (Ulmaceae). These are small secondary trees characteristic of man-made vegetation, in both East and West Malesia. The pollen is abundantly produced and apparently spreads widely, so is presumably of high relative export.

Urticaceae/Moraceae. The pollen of these is not usually distinguished. Both families are most abundant in secondary vegetation, although both are well represented in primary forest. An increase in the pollen type probably indicates forest disturbance. Urticaceous shrubs are preserved in New Guinea to provide fibres.

Vernonia (Compositae). *V. arborea* is a tree, occurring in West Malesia from sea level up to 2400 m, but most common in mountain forest. Some other *Vernonia* species are shrubs or climbers.

Bibliography

ADAMS, C. G. and AGER, D.V. (1967). *Aspects of Tethyan Biogeography: a symposium, Systematics Assocn. Publn.* 7, London.

ALDRICH-BLAKE, R. N. (1932). 'On the fixation of atmospheric nitrogen by bacteria living symbiotically in root nodules of *Casuarina equisetifolia.' Oxf. For. Mem.* 14, 1–20.

ALEVA, G. J. J. (1973). 'A contribution to the geology of part of the Indonesian tinbelt: the sea areas between Singkap and Bangka islands and around the Karimata islands.' (Proc. Reg. Conf. Geol. S. E. Asia.) *Bull. geol. Soc. Malaysia* 6, 257–271.

AMBROSE, W. (1976). 'Intrinsic hydration dating of obsidian.' *Advances in Obsidian Glass Studies,* (ed. R. E. Taylor) Noyes Press, New York.

ANDEL, T. H. van, HEATH, G. R., MOORE, T. C. and McGEARY, D. F. R. (1967). 'Late Quaternary history, climate and oceanography of the Timor Sea, north-western Australia.' *Am. J. Sci.* 265, 737–758.

ANDEL, T. H. van and VEEVERS, J. J. (1967). 'Morphology and sediments of the Timor Sea.' *Commonwealth of Australia, Bureau of Mineral Resources, Geology and Geophysics Bull.* 83.

ANDERSON, J. A. R. (1958). 'Observations on the ecology of the peat swamp forests of Sarawak and Brunei.' *Proceedings of the Symposium on Humid Tropics vegetation (Tjiawi) Indonesia, Dec. 1958.* 141–149, UNESCO.

ANDERSON, J. A. R. (1960). 'Research into the effects of shifting cultivation in Sarawak.' *Symposium on the impact of man on humid tropics vegetation, Goroka, Territory of Papua and New Guinea, Sept. 1960.* 203–206, UNESCO.

ANDERSON, J. A. R. (1961). 'The destruction of *Shorea albida* forest by an unidentified insect.' *Emp. For. Rev.* 40, 19–29.

ANDERSON, J. A. R. (1963). 'The flora of the peat swamp forests of Sarawak and Brunei, including a catalogue of all recorded species of flowering plants, ferns and fern allies.' *Gdns' Bulletin, Singapore* 20, 131–145.

ANDERSON, J. A. R. (1964). 'The structure and develop-
ment of the peat swamps of Sarawak and Brunei.' *J. trop. Geogr.* 18, 7–16.

ANDERSON, J. A. R. and MULLER, J. (1975). 'Palynological study of a Holocene peat and a Miocene coal deposit from N. W. Borneo.' *Rev. Palaeobot. & Palynol.* 19, 291–351.

ANON. (Ed.) (1963) *Simposio sobre o Cerrado.* Universidade de São Paulo, São Paulo, Brazil, 424 pp.

ASHTON, P. S. (1964). 'Ecological Studies in the mixed Dipterocarp forests of Brunei State.' *Oxf. For. Mem.* 25, 1–75.

ASHTON, P. S. (1969). 'Speciation among tropical forest trees: some deductions in the light of recent evidence.' *Biol. J. Linn. Soc.* 1, 155–196.

ASHTON, P. S. (1972). 'The Quaternary geomorphological history of western Malesia and lowland forest phytogeography.' *The Quaternary Era in Malesia. Transactions of the Second Aberdeen-Hull Symposium on Malesian Ecology, Aberdeen 1971,* (ed. P. and M. Ashton), 35–49, University of Hull, Dept. of Geography, Miscellaneous Series No 13, Hull, England.

ASPREY, G. F. and ROBBINS, R. G. (1953). 'The vegetation of Jamaica.' *Ecol. Monogr.* 23, 359–412.

AUBRÉVILLE, A. (1949). *Climats, Forêts et désertification de l'Afrique tropicale,* Société d'Editions Géographiques, Martimes et Coloniales, Paris.

AUBRÉVILLE, A. (1950–51). 'Le concept d'Association dans la forêt dense équatoriale de la basse Côte d'Ivoire.' *Bull. Soc. bot. Fr.* 97, 145–158.

AXELROD, D. I. (1959). 'Poleward migration of early angiosperm flora.' *Science, N. Y.* 130, 203–207.

AXELROD, D. I. (1963). 'Fossil floras suggest stable, not drifting continents.' *J. geophys. Res.* 68, 3257–3263.

AXELROD, D. I. and BAILEY, H. P. (1969). 'Palaeotemperature analysis of Tertiary floras.' *Palaeogeogr. Palaeoclimatol. Palaeoecol.* 6, 163–195.

BACKER, C. A. (1929). *The problem of Krakatoa as seen by a botanist,* 229 pp and maps. Welterreden (Java) and The Hague.

BAKKER, E. M. van ZINDEREN (1962). 'A Late-Glacial and Post-Glacial climatic correlation between East Africa and Europe.' *Nature, Lond.* **194**, 201–203.

BAKKER, E. M. van ZINDEREN (1964). 'A pollen diagram from equatorial Africa, Cherangani, Kenya.' *Geologie Mijnb.* **43**, 123–128.

BAKKER, E. M. van ZINDEREN and COETZEE, J. A. (1972). 'A re-appraisal of Late-Quaternary climatic evidence from tropical Africa.' *Palaeoecol. Af. & Surround. Isles Antarct.* **7**, 151–181.

BALGOOY, M. M. J. van (1971). 'Plant geography of the Pacific.' *Blumea Suppl. to vol* **6**, 222 pp.

BANCROFT, H. (1932). 'Some fossil dicotyledonous woods from the Miocene (?) beds of East Africa.' *Ann. Bot.* **46**, 745–767.

BANCROFT, H. (1933a). 'On certain fossil plants from East Africa.' *Ann. Bot.* **47**, 915–920.

BANCROFT, H. (1933b). 'A contribution to the geological history of the Dipterocarpaceae.' *Geol. För. Stockh. Förh.* **55**, 59–100.

BANDY, O. L. (1968). 'Cycles in Neogene paleoceanography and eustatic changes.' *Palaeogeogr. Palaeoclimatol. Palaeoecol.* **5**, 63–75.

BARBETTI, M. and ALLEN, H. (1972). 'Prehistoric man at Lake Mungo, Australia by 32 000 years B.P.' *Nature, Lond.* **240**, 46–48.

BARGHOORN, E. S., WOLFE, M. K. and CLISBY, K. H. (1954). 'Fossil maize from the valley of Mexico.' *Harvard Univ. Bot. Mus. Leafl.* **16**, 229–240.

BARNARD, P. D. W. (1973). 'Mesozoic floras.' *Organisms and Continents Through Time, Special papers in Palaeontology* **12**, 175–187, *Systematics Assocn. Publ.* **9**, London.

BARTLETT, A. S. and BARGHOORN, E. S. (1973). 'Phytogeographic history of the Isthmus of Panama during the past 12 000 years (A history of vegetation, climate, and sea-level change).' *Vegetation and Vegetational History of Northern Latin America, a Symposium.* (ed. A. Graham). 203–299, Elsevier, Amsterdam & New York.

BARTLETT, A. S., BARGHOORN, E. S. and BERGER, R. (1969). 'Fossil maize from Panama.' *Science, N. Y.* **165**, 389–390.

BARTLETT, H. H. (1956). 'Fire, primitive agriculture and grazing in the tropics.' *Man's role in changing the face of the Earth* (ed. W. L. Thomas, C. O. Sauer, M. Bates and L. Mumford). 692–720, University of Chicago Press, Chicago.

BAUR, G. N. (1964). *The ecological basis of rainforest management.* Sydney

BEARD, J. S. (1946). 'The natural vegetation of Trinidad.' *Oxf. For. Mem.* **20**, 1–152.

BEARD, J. S. (1953). 'The savanna vegetation of northern tropical America.' *Ecol. Monogr.* **23**, 149–215.

BEMMELEN, R. W. van (1949). *The geology of Indonesia,* Vols Ia, Ib & II, Govt Printing Office, The Hague.

BERRY, E. W. (1919). 'The fossil higher plants from the Canal Zone.' *Bull. U. S. natn. Mus.* **103**, 15–44.

BERRY, E. W. (1925). 'The Tertiary flora of the island of Trinidad.' *Johns Hopkins Univ. Stud. Geol.* **6**, 71–162.

BERRY, E. W. (1929). 'Fossil fruits in the Ancon Sandstone of Ecuador.' *J. Paleont.* **3**, 298–301.

BERRY, E. W. (1936). 'Tertiary plants from Colombia.' *Bull. Torrey bot. Club* **63**, 53–66.

BERRY, E. W. (1937a). 'A flora from the forest clay of Trinidad, B. W. I.' *Johns Hopkins Univ. Stud. Geol.* **12**, 51–68 and Plates X–XIII.

BERRY, E. W. (1937b). 'A Late Tertiary Flora from Trinidad.' *Johns Hopkins Univ. Stud. Geol.* **12**, 69–79 and Plate XIV.

BERRY, E. W. (1939). 'Eocene plants from a well core in Venezuela.' *Johns Hopkins Univ. Stud. Geol.* **13**, 157–164.

BERRY, E. W. (1942). 'Mesozoic and Cenozoic plants of South America, Central America and the Antilles.' *Proc. 8th Am. Sci. Congr. Washington D.C., 1940* **4**, 365–373.

BEUCHER, F. (1967). 'Une flore d'âge ougartien (seconde partie du Quaternaire moyen) dans les monts d'Ougarta Sahara nord-occidental.' *Rev. Palaeobot. & Palynol,* **2**, 291–300.

BISWAS, B. (1973). 'Quaternary changes in sea-level in the South China Sea.' (Proc. Reg. Conf. Geol. S. E. Asia), *Bull. geol. Soc. Malaysia* **6**, 229–255.

BOLTENHAGEN, E. (1976). 'La Microflore Sénonienne du Gabon.' *Revue de Micropaléontologie,* **18**, 191–199.

BONNEFILLE, R. (1969). 'Analyse pollinique d'un sediment récent: vases actuelles de la rivière Aouache (Ethiopie).' *Pollen Spores* **11**, 7–16.

BONNEFILLE, R. (1972). *Associations polliniques actuelles et Quaternaires en Ethiopie (Vallées de l'Awash et de l'Omo),* Thesis submitted to Université de Paris VI for Dr ès Sciences Naturelles, 2 vols. 513 pp.

BONNEFILLE, R. (1973). 'Nouvelles recherches en Ethiopie.' *Le Quaternaire. Geodynamique, stratigraphie et environnement. Travaux Français Récents.* 182–187 Comité National Français de l'INQUA.

BONNEFILLE, R. (in press). 'Vegetation history of savanna in East Africa during the Plio-Pleistocene.' *Proc. IV International Palynological Conference, Lucknow, Jan. 1977.*

BOWDLER, S. (1977). 'The coastal colonisation of Australia.' *Sunda and Sahul: Prehistoric study in southeast Asia, Melanesia and Australia,* (ed. J. Allen, J. Golson and R. Jones) Academic Press, London.

BOWLER, J. M., HOPE, G. S., JENNINGS, J. N., SINGH, G. and WALKER, D. (1976). 'Late Quaternary climates of Australia and New Guinea.' *Quaternary Res.* **6**, 359–394.

BRASS, L. J. (1941). 'The 1938–39 Expedition to the Snow Mountains, Netherlands New Guinea.' *J. Arnold Arbor.* **22**, 271–342.

BRASS, L. J. (1964). 'Results of the Archbold Expeditions No. 86. Summary of the Sixth Archbold Expedition to New Guinea (1959).' *Bull. Am. Mus. nat. Hist.* **127**, 145–216.

BRAUN-BLANQUET, J. (1932). *Plant Sociology; the study of plant communities,* 439pp, McGraw-Hill, New York & London.

BRENNER, G. J. (1968). 'Middle Cretaceous spores and pollen from northeastern Peru.' *Pollen Spores* **10**, 341–383.

BROOKFIELD, H. C. (1964). 'The Ecology of highland settlement: some suggestions.' *Am. Anthrop.* **66**, 20–38.

BROOKFIELD, H. C. and BROWN, P. (1963). *Struggle for land: agriculture and group territories among the Chimbu of the New Guinea highlands,* Oxford University Press, Melbourne.

BROOKS, J. and SHAW, G. (1968). 'Identity of sporopollenin with older kerogen and new evidence for the possible biological source of chemicals in sedimentary rocks.' *Nature, Lond.* **220**, 678–679.

BROWN, K. S. Jr. (1972). 'The heliconians of Brazil (Lepidoptera: Nymphalidae). Part III Ecology and biology of *Heliconius nattereri,* a key primitive species near extinction, and comments on the evolutionary development of *Heliconius* and *Eueides.' Zoologica, N. Y.* **57**, 41–69.

BROWN, M. and POWELL, J. M. (1974). 'Frost and drought in the Highlands of Papua New Guinea.' *J. trop. Geogr.* **38**, 1–6.

BRUNIG, E. F. (1971). 'On the ecological significance of drought in the Equatorial wet evergreen (rain) forest of Sarawak (Borneo).' *The Water Relations of Malesian Forests. Transactions of the First Aberdeen - Hull Symposium on Malesian Ecology, Hull 1970,* (ed. J. R. Flenley) 66–88, University of Hull, Dept. of Geography, Miscellaneous Series No. 11, Hull, England.

BRYAN, A. L. (1973). 'Paleoenvironments and cultural diversity in Late Pleistocene South America.' *Quaternary Res.* **3**, 237–256.

BULMER, S. (1975). 'Settlement and economy in prehistoric Papua New Guinea: a review of the archaeological evidence.' *J. Soc. Océan.* **46**, 31, 7–75.

BURTON, C. K. (1970). 'Palaeotectonic status of the Malay Peninsula.' *Palaeogeogr. Palaeoclimatol. Palaeoecol.* **7**, 51–60.

BUTZER, K. W., ISAAC, G. L., RICHARDSON, J. L. and WASHBOURN-KAMAU, C. (1972). 'Radiocarbon Dating of East African Lake Levels.' *Science, N. Y.* **175**, 1069–1076.

CAMPO, M. van (1967). 'Etude de la Désertification du Massif du Hoggar par les méthodes de l'analyse pollinique.' *Rev. Palaeobot. & Palynol.* **2**, 281–289.

CAMPO, M. van. GUINET, P., COHEN, J. and DUTIL, P. (1967). 'Contribution à l'étude du peuplement végétal quaternaire des montagnes sahariennes. III Flore de l'Oued Outoul (Hoggar).' *Pollen Spores* **9**, 107–120.

CARR, S. G. M. (1972). 'Problems of the geography of the tropical Eucalypts.' *Bridge and Barrier: the Natural and Cultural History of Torres Strait* (ed. D. Walker) 153–181, Australian National University, Research School of Pacific Studies, Department of Biogeography and Geomorphology, Publication BG/3. Canberra.

CHANEY, R. W. (1933). 'A Tertiary flora from Uganda.' *J. Geol.* **41**, 702–709.

CHANEY, R. W. (1947). 'Tertiary centres and migration routes.' *Ecol. Monogr.* **17**, 139–148.

CHANEY, R. W. and SANBORN, E. J. (1933). 'The Goshen flora of west central Oregon.' *Publs Carnegie Instn* **439**, 103.

CHAPPELL, J. (1973). 'Astronomical theory of climatic change: status and problem.' *Quaternary Res.* **3**, 221–236.

CHESTERS, K. I. M. (1957). 'The Miocene Flora of Rusinga Island, Lake Victoria, Kenya.' *Palaeontographica* **101**, 30–71.

CHESTERS, K. I. M., GNAUCK, F. R. and HUGHES, N. F. (1967). 'Angiospermae.' *The Fossil Record*, (ed. W. B. Harland *et al*) Chapter 7, Geological Society London.

CLARK, J. D. (1962). 'The spread of food production in sub-Saharan Africa.' *J. Afr. Hist.* **3**, 211–228.

CLARK, J. D. (1967). *Atlas of African Prehistory*, 62pp, University of Chicago Press, Chicago.

CLARK, J. D. (1970). *The Prehistory of Africa*, 302pp, Thames & Hudson, London.

CLASON, E. W. (1935). 'The vegetation of the Upper-Badak region of Mount Kelut (East Java).' *Bull. Jard. bot. Buitenz. III*, **13**, 509–518.

CLEMENTS, F. E. (1916). 'Plant succession; an analysis of the development of vegetation.' *Publs Carnegie Instn* **242**.

COE, M. J. (1967). *The ecology of the Alpine Zone of Mt. Kenya. Monographiae biol.* **17**, 136pp, W. Junk, The Hague.

COETZEE, J. A. (1967). 'Pollen analytical studies in East and Southern Africa.' *Palaeoecol. Af. & Surround. Isles Antarct.* **3**, 1–146.

COHEN, J. (1970). 'Analyse pollinique.' *Quelques formations lacustres et fluviatiles, au Tibesti (N. du Tchad).* (By J. Maley, J. Cohen, A. Faure, P. Rognan, P. M. Vincen) *Cahier ORSTOM, Sér. Geol.* II(1), 146–150.

COLE, M. M. (1960). 'Cerrado, caatinga and pantanal: the distribution and origin of the savanna vegetation of Brazil.' *Geogrl. J.* **126**, 168–179.

COLINVAUX, P. A. (1968). 'Reconnaissance and Chemistry of the Lakes and Bogs of the Galapagos Islands.' *Nature, Lond.* **219**, 590–594.

COLINVAUX, P. A. (1972). 'Climate and the Galapagos Islands.' *Nature, Lond.* **240**, 17–20.

CONKLIN, H. C. (1963). 'The Oceanian - African hypotheses and the migration of the sweet potato.' *Plants and the migration of Pacific peoples. Symposium of the 10th Pacific Science Congress, Hawaii, 1961*, (ed. J. Barrau) 129–136, Bishop Museum Press, Honolulu.

COTTON, A. D. (1944). 'The megaphytic habit in the tree Senecios and other genera.' *Proc. Linn. Soc. Lond.* **156**, 158–168.

COWGILL, U. M., GOULDEN, C. E., HUTCHINSON, G. E., PATRICK, R., RACEK, A. A. and TSUKADA, M. (1966). 'The history of Laguna de Petenxil, a small lake in northern Guatemala.' *Mem. Conn. Acad. Arts Sci.* **17**, 1–126.

CRANWELL, L. M. (1963). 'Nothofagus - Living and Fossil.' *Pacific Basin Biogeography. Symposium of the 10th Pacific Science Congress, Hawaii 1961*, (ed. J. L. Gressitt) 387–400.

CRUSZ, H. (1973). 'Nature Conservation in Sri Lanka (Ceylon).' *Biol. Conserv.* **5**, 199–208.

CUATRECASAS, J. (1958). 'Aspectos de la Vegetacion Natural de Colombia.' *Revta Acad. colomb. Cienc. exact. fis. nat.* **10**, 221–264.

DALEY, B. (1972). 'Some problems concerning the Early Tertiary climate of Southern Britain.' *Palaeogeogr. Palaeoclimatol. Palaeoecol.* **11**, 177–190.

DARWIN, C. (1859). *The origin of species by means of natural selection*, 432pp, John Murray, London.

DIAMOND, J. M. (1975). 'The island dilemma: lessons of modern biogeographic studies for the design of natural reserves.' *Biol. Conserv.* **7**, 129–146.

DICKENSON, D. and PORTER, J. (1972). 'Is this the place where man first came to grow rice?' *The Asian*, 20–26 August 1972, p 8. (newspaper article).

DILCHER, D. L. (1973). 'A Palaeoclimatic interpretation of the Eocene floras of Southeastern North America.' *Vegetation and Vegetational History of Northern Latin America*, (ed. A. Graham.) 36–59, Elsevier, Amsterdam & New York.

DOCTERS van LEEUWEN, W. M. (1936). 'Krakatau, 1883 to 1933. A. Botany.' *Ann. bot. Gdn. Buitenz.* **46–47**, 1–506.

DOLIANITI, E. (1955). 'Frutos de Nipa no Paleoceno de Pernambuco, Brasil,' *Bras. Dep. Nacl. Prod. Min., Div. Geol. Min, Bol.* **158**, 1–36.

DOUGLAS, I. (1967). 'Man, vegetation and the sediment yields of rivers.' *Nature, Lond.* **215**, 925–928.

DOUGLAS, I. (1971). 'Aspects of the water balance of catchments in the Main Range near Kuala Lumpur.' *The Water Relations of Malesian Forests, Transactions of the First Aberdeen-Hull Symposium on Malesian Ecology*, (ed. J. R. Flenley) 23–35, University of Hull, Dept. of Geography, Miscellaneous Series No 11, Hull, England.

DOYLE, J. A., van CAMPO, M. and LUGARDON, B. (1975). 'Observations on exine structure of *Eucommiidites* and Lower Cretaceous angiosperm pollen.' *Pollen Spores* **17**, 429–486.

EDEN, M. J. (1974). 'Palaeoclimatic influences and the development of savanna in southern Venezuela.' *J. Biogeogr.* **1**, 95–109.

ELSIK, W. C. (1974). '*Nothofagus* in N. America.' *Pollen Spores* **16**, 285–299.

EMILIANI, C. (1955). 'Pleistocene Temperatures.' *J. Geol.* **63**, 538–578.

EMILIANI, C. (1971). 'The amplitude of Pleistocene climatic cycles at low latitudes and the isotopic composition of glacial ice.' *The Late Cenozoic Glacial Ages*, (ed. K. K. Turekian) 183–197, Yale University Press, New Haven.

FAEGRI, K. (1966). 'Some problems of representativity in pollen analysis.' *Palaeobotanist* **15**, 135–140.

FAEGRI, K. and IVERSEN, J. (1975). *Textbook of pollen analysis*, 3rd Edn., 295pp, Blackwell, Oxford.

FEDOROV, AN. A. (1966). 'The Structure of Tropical Rain Forest and Speciation in the Humid Tropics.' *J. Ecol.* **54**, 1–11.

FERRI, M. G. (Ed.) (1971). *III Simposio sobre o Cerrado*. Universidade de São Paulo, São Paulo, Brazil. 239 pp.

FITCH, F. H. (1952). 'The geology and mineral resources of the neighbourhood of Kuantan, Pahang.' *Mem. geol. Surv. Dep. Fed. Malaya* **6**, 1–143.

FLENLEY, J. R. (1967). *The present and former vegetation of the Wabag region of New Guinea*, Ph. D. thesis, Australian National University, Canberra, 379pp.

FLENLEY, J. R. (1969). 'The vegetation of the Wabag region, New Guinea Highlands: a numerical study.' *J. Ecol.* **57**, 465–490.

FLENLEY, J. R. (1972). 'Evidence of Quaternary vegetational change in New Guinea.' *The Quaternary Era in Malesia, Transactions of the Second Aberdeen-Hull Symposium on Malesian Ecology, Aberdeen 1971,* (ed. P. and M. Ashton) 99–109, University of Hull, Department of Geography, Miscellaneous Series, No 13, Hull, England.

FLENLEY, J. R. (1973). 'The use of modern pollen rain samples in the study of the vegetational history of tropical regions.' *Quaternary Plant Ecology, The 14th Symposium of the British Ecological Society, Mar. 1972,* (ed. H. J. B. Birks and R. G. West) 131–141, Blackwell, Oxford.

FLENLEY, J. R. (in press). 'Conclusion: the Quaternary history of the tropical rain forest and other vegetation of tropical mountains.' *Proc. IV International Palynological Conference, Lucknow, Jan. 1977.*

FLENLEY, J. R. and MORLEY, R. J. (1978). 'A minimum age for the deglaciation of Mt Kinabalu, East Malaysia.' *Modern Quaternary Research in South-East Asia* 4, 57–61.

FLENLEY, J. R., MORLEY, R. J. and MALONEY, B. K. (1976). *Investigation of Quaternary Vegetational History in South-East Asia,* 39pp, Final report to the Natural Environment Research Council of Great Britain.

FLINT, R. F. (1971). *Glacial and Quaternary Geology,* New York, J. Wiley.

FLORIN, R. (1963). 'The distribution of conifer and taxad genera in time and space.' *Acta Horti Bergiani* 20, No. 4, 1–278.

FONAROFF, L. S. (1974). 'Urbanization, birds and ecological change in northwestern Trinidad.' *Biol. Conserv.* 6, 258–262.

GALINAT, W. C. (1971). 'The origin of maize.' *A. Rev. Genet.* 5, 447–478.

GEEL, B. van and HAMMEN, T. van der, (1973). 'Upper Quaternary vegetational and climatic sequence of the Fuquene area (Eastern cordillera, Colombia).' *Palaeogeogr. Palaeoclimatol. Palaeoecol.* 14, 9–92.

GEIKIE, Sir A. (1923). *Text book of Geology.* 4th Edition, 2 vols, 1472pp, Macmillan, London.

GERMERAAD, J. H., HOPPING, C. A. and MULLER, J. (1968). 'Palynology of Tertiary Sediments from Tropical Areas.' *Rev. Palaeobot. & Palynol.* 6, 189–348.

GILLETT, J. B. (1962). 'Pest pressure, an underestimated factor in evolution.' *Taxonomy and Geography.* 37–46, *Systematics Assocn Publn* 4, London.

GLEASON, H. A. (1926). 'The individualistic concept of the plant association.' *Bull. Torrey bot. Club* 53, 7–26.

GLOVER, I. C. (1969). 'Radiocarbon dates from Portuguese Timor.' *Archaeol. & Phys. Anthropol. Oceania* 4, 107–112.

GLOVER, I. C. (1973). 'Late Stone Age traditions in South-East Asia.' *Papers from the First International Conference of South Asian Archaeologists held in the University of Cambridge,* (ed. N. Hammond) 51–65.

GODWIN, H. (1956). *The History of the British Flora; a factual basis for phytogeography,* 383pp, Cambridge University Press, Cambridge.

GOLSON, J. (1977). 'No room at the top: agricultural intensification in the New Guinea Highlands.' *Sunda and Sahul: Prehistoric study in southeast Asia, Melanesia and Australia,* (ed. J. Allen, J. Golson, and R. Jones) Academic Press, London.

GOLSON, J. and HUGHES, P. J. (1976). 'The appearance of plant and animal domestication in New Guinea.' *IXth Congress, Union International des Sciences Prehistoriques et Protohistoriques, Nice, Sept. 1976.*

GONZALEZ, E., HAMMEN, T. van der and FLINT, R. F. (1966). 'Late Quaternary glacial and vegetational sequence in Valle de Lagunillas, Sierra Nevada del Cocuy, Colombia.' *Leid. geol. Meded.* 32, 157–182.

GOOD, R. (1947). *The geography of the flowering plants,* 1st edition, 403pp, Longman, London.

GOOD, R. (1956). *Features of evolution in the flowering plants,* 405pp, Longman, London.

GORMAN, C. (1972). 'Excavations at Spirit Cave, North Thailand, 1966: some interim interpretations.' *Asian Perspectives* 13, 79–107.

GRAHAM, A. (1973). 'History of the Arborescent Temperate Element in the Northern Latin American Biota.' *Vegetation and Vegetational History of Northern Latin America,* (ed. A. Graham) 301–314, Elsevier, Amsterdam & New York.

GREENWAY, P. J. (1965). 'The vegetation and flora of Mt. Kilimanjaro.' *Tanganyika Notes & Records* 64, 97–108.

GREIG-SMITH, P. (1952). 'Ecological observations on degraded and secondary forest in Trinidad, British West Indies. I. General features of the vegetation.' *J. Ecol.* 40, 283–315.

GRUBB, P. J. (1971). 'Interpretation of the 'Massenerhebung' Effect on Tropical Mountains.' *Nature, Lond.* 229, 44–45.

GRUBB, P. J. (1974). 'Factors controlling the distribution of forest-types on tropical mountains: new facts and a new perspective.' *Altitudinal Zonation in Malesia, Transactions of the Third Aberdeen-Hull Symposium on Malesian Ecology, Hull 1973,* (ed. J. R. Flenley) 13–46, University of Hull, Dept. of Geography, Miscellaneous Series No 16, Hull, England.

GRUBB, P. J. and WHITMORE, T. C. (1966). 'A comparison of montane and lowland rain forest in Ecuador. 2. The climate and its effects on the distribution and physiognomy of the forests.' *J. Ecol.* 54, 303–333.

GUPTA, H. P. (1971). 'Quaternary vegetational history of Ootacamund, Nilgiris, South India. I. Kakathope and Rees Corner.' *Palaeobotanist* 20, 74–90.

HAFFER, J. (1969). 'Speciation in Amazonian forest birds.' *Science, N. Y.* 165, 131–137.

HAFSTEN, U. (1960). 'Pleistocene development of vegetation and climate in Tristan da Cunha and Gough Island.' *Årbok Univ. Bergen, Mat-Naturv. Serie* 20, 1–48.

HAILE, N. S. (1971). 'Quaternary shorelines in West Malaysia and adjacent parts of the Sunda Shelf.' *Quaternaria* 15, 333–343.

HAMANN, O. (1975). 'Vegetational changes in the Galapagos Islands during the period 1966–1973.' *Biol. Conserv.* 7, 37–59.

HAMILTON, A. C. (1968). 'Some plant fossils from Bukwa.' *Uganda J.* 32 (2), 157–164.

HAMILTON, A. C. (1972). 'The interpretation of pollen diagrams from Highland Uganda.' *Palaeoecol. Af. & Surround. Isles Antarct.* 7, 45–149.

HAMILTON, A. C. (1974). 'The history of the vegetation.' *East African Vegetation* (by E. M. Lind and M. E. S. Morrison) 188–209, Longman, London.

HAMMEN, T. van der (1961a). 'Upper Cretaceous and Tertiary climatic periodicities and their causes.' *Ann. N.Y. Acad. Sci.* 95, 440–448.

HAMMEN, T. van der (1961b). 'Deposicion Reciente de Polen Atmosferico en la Sabana de Bogota y Alrededores.' *Boln. geol., Bogota* 7, No. 1–3, 183–194.

HAMMEN, T. van der (1961c). 'The Quaternary climatic changes of northern South America.' *Ann. N. Y. Acad. Sci.* 95, 676–683.

HAMMEN, T. van der (1962). 'Palinologia de la Region de "Laguna de los Bobos". Historia de su clima, vegetacion y agricultura durante los ultimos 5000 años.' *Revta Acad. colomb. Cienc. exact. fis. nat.* 11 (44), 359–361.

HAMMEN, T. van der (1963). 'A palynological study on the Quaternary of British Guiana.' *Leid. geol. Meded.* 29, 125–180.

HAMMEN, T. van der (1974). 'The Pleistocene changes of vegetation and climate in tropical South America.' *J. Biogeogr.* 1, 3–26.

HAMMEN, T. van der and GONZALEZ, E. (1960). 'Upper Pleistocene and Holocene climate and vegetation of the 'Sabana de Bogota' (Colombia, South America).' *Leid. geol. Meded.* 25, 261–315.

HAMMEN, T. van der and GONZALEZ, E. (1964). 'A pollen diagram from the Quaternary of the Sabana de Bogota (Colombia) and its significance for the geology of the Northern Andes.' *Geologie Mijnb.* 43, 113–117.

HAMMEN, T. van der and GONZALEZ, E. (1965a). 'A pollen diagram from "Laguna de la Herrera" (Sabana de Bogota).' *Leid. geol. Meded.* 32, 183–191.

HAMMEN, T. van der and GONZALEZ, E. (1965b). 'A Late-glacial and Holocene pollen diagram from Cienaga del Visitador (Dept. Boyaca, Colombia).' *Leid. geol. Meded.* 32, 193–201.

HAMMEN, T. van der, WERNER, J. H. and DOMMELEN, H. van (1973). 'Palynological record of the upheaval of the Northern Andes: a study of the Pliocene and Lower Quaternary of the Colombian Eastern Cordillera and the early evolution of its High Andean biota.' *Rev. Palaeobot. & Palynol.* 16, 1–122.

HARLAND, W. B., SMITH, G. A. and WILCOCK, B. (1964). (Eds). 'The Phanerozoic Time Scale. A symposium dedicated to Professor A. Holmes.' *Suppl. Quart. J. geol. Soc. London* 120.

HARRISSON, T. (1972). 'The prehistory of Borneo.' *Asian Perspectives* 13, 17–45.

HASTENRATH, S. (1968). 'Certain aspects of the three-dimensional distribution of climate and vegetation belts in the mountains of central America and southern Mexico.' *Colloquium geogr.* 9, 122–130.

HAVEL, J. J. (1971). 'The *Araucaria* forests of New Guinea and their regenerative capacity.' *J. Ecol.* 59, 203–214.

HAYNES, V. (1974). 'Paleoenvironments and cultural diversity in late Pleistocene South America: a reply to A. L. Bryan.' *Quaternary Res.* 4, 378–382.

HAYS, J. D., IMBRIE, J. and SHACKLETON, N. J. (1976). 'Variations in the Earth's Orbit: Pacemaker of the Ice Ages.' *Science, N. Y.* 194, 1121–1132.

HEDBERG, O. (1951). 'Vegetation belts of the East African Mountains.' *Svensk bot. Tidskr.* 45, 140–202.

HEDBERG, O. (1954). 'A pollen-analytical reconnaissance in tropical East Africa.' *Oikos* 5, 137–166.

HEDBERG, O. (1957). 'Afroalpine vascular plants.' *Symb. Bot. Upsal.* 15, 1–411.

HEDBERG, O. (1964). 'Features of Afroalpine Plant Ecology.' *Acta phytogeogr. suec.* 49, 1–144.

HEDEGART, T. (1976). 'Breeding systems, variation and genetic improvement of Teak.' (*Tectona grandis* L.f.). *Tropical Trees. Variation, Breeding and Conservation, Linnean Society Symposium Series, No. 2,* (ed. J. Burley and B. T. Styles) 109–123, Academic Press, London.

HEEKEREN, H. R. van (1972). *The Stone Age of Indonesia.* 2nd Edn. with a contribution by R. P. Soejono. Verhandelingen van het Koninklijk Instituut voor Taal - Land - en Volkenkunde No. 61, 247 pp. and 105 plates, Martinus Nijhoff, The Hague.

HERNGREEN, G. F. W. (1973). 'Palynology of Albian-Cenomanian Strata of Borehole 1-QS-1-MA, State of Maranhao, Brazil.' *Pollen Spores* 15, 515–555.

HERNGREEN, G. F. W. (1975). 'Palynology of Middle and Upper Cretaceous strata in Brazil.' *Meded. Rijks geol. Dienst., N. S.* 26, No. 3, 39–91.

HICKEY, L. J. and WOLFE, J. A. (1975). 'The Bases of Angiosperm Phylogeny: Vegetative morphology.' *Ann. Mo. bot. Gdn* 62, 538–589.

HILLS, T. L. (1965). 'Savannas: a review of a major research problem in tropical geography.' *Can. Geogr.* 9, part 4, 216–228.

HOLMES, C. H. (1951). 'The grass, fern and savannah lands of Ceylon, their nature and ecological significance.' *Imperial Forestry Inst., Oxford Inst. Paper No* 28, 95pp.

HOLTTUM, R. E. (1954). '*Adinandra belukar.' J. trop. Geogr.* 3, 27–32.

HOOP, A. N. J. Th. à Th. van der (1940). 'A prehistoric site near the Lake of Kerinchi (Sumatra).' *Proc. 3rd Congr. of Prehistorians of the Far East 1938, Singapore,* (ed. F. N. Chasen and M. W. F. Tweedie) 200–204.

HOOPER, M. D. (1971). 'The size and surroundings of nature reserves.' *The scientific management of animal and plant communities for conservation: a symposium. Eleventh Symposium of the British Ecological Society, Norwich 1970,* (ed. E. Duffy and A. S. Watt) 555–561, Blackwell, Oxford.

HOPE, G. S. (1973). *The vegetation history of Mt Wilhelm, Papua New Guinea.* Ph.D. thesis, Australian National University, Canberra, 460pp.

HOPE, G. S. (1976a). 'The vegetational history of Mt Wilhelm, Papua New Guinea.' *J. Ecol.* 64, 627–663.

HOPE, G. S. (1976b). 'Vegetation' *The Equatorial Glaciers of New Guinea* (ed. G. S. Hope, J. A. Peterson and U. Radok) Chapter 8, A. A. Balkema, Rotterdam.

HOPE, G. S. and PETERSON, J. A. (1975). 'Glaciation and vegetation in the High New Guinea Mountains.' *Bull. R. Soc. N. Z.* 13, 155–162.

HOPE, G. S. and PETERSON, J. A. (1976). 'Palaeoenvironments.' *The Equatorial Glaciers of New Guinea* (ed. G. S. Hope, J. A. Peterson and U. Radok), Chapter 9, A. A. Balkema, Rotterdam.

HOPLEY, D. (1970). *The geomorphology of the Burdekin Delta, North Queensland.* Dept. Geography, James Cook University of North Queensland, Monograph Series No 1, 66pp.

HOUVENAGHEL, G. T. (1974). 'Equatorial undercurrent and climate in the Galapagos Islands.' *Nature, Lond.* 250, 565–566.

HOWELL, W. W. (1973). *The Pacific Islanders.* 299pp, Weidenfeld & Nicolson, London.

HUECK, K. (1966). *Die Wälder Sudamerikas.* 422pp, Gustav Fischer Verlag, Stuttgart.

HUECK, K. and SEIBERT, P. (1972). 'Vegetationskarte von Sudamerika. Mapa de la Vegetacion de America del Sur.' *Vegetations mono-graphien der Einzelnen Grossraume* 11a, 71 pp. Gustav Fischer Verlag, Stuttgart.

HUERTOS, G. C. and HAMMEN, T. van der (1953). 'Un possible banano (*Musa*) fosil del Cretaco de Colombia.' *Revta Acad. colomb. Cienc. exact. fis. nat.* 9, 115–116.

HUGHES, N. F. (1961). 'Fossil evidence and angiosperm ancestry.' *Sci. Prog., Lond.* 49, No. 193, 84–102.

HUGHES, N. F. (1976). *Palaeobiology of Angiosperm Origins,* 242pp, Cambridge University Press, Cambridge.

HUMBOLDT, A. von (1852). *Personal narrative of travels to the equinoctial regions of America,* 2, Transl. T. Ross, London.

HUMMEL, K. (1931). Sedimente indonesischer Süsswasserseen. *Arch. Hydrobiol. Suppl. Bd.* 8, 615–676.

HUTCHINSON, G. E., PATRICK, R. and DEEVEY, E. S. (1956). 'Sediments of Lake Patzcuaro, Michoacan, Mexico.' *Bull. geol. Soc. Am.* 67, 1491–1504.

HYNES, R. A. (1974). 'Altitudinal zonation in New Guinea *Nothofagus* forests.' *Altitudinal Zonation in Malesia, Transactions of the Third Aberdeen-Hull Symposium on Malesian Ecology, Hull 1973,* (ed. J. R. Flenley) 75–120, University of Hull, Dept. of Geography, Miscellaneous Series No 16, Hull, England.

IRWIN, H. and BARGHOORN, E. S. (1965). 'Identification of the pollen of maize, teosinte and *Tripsacum* by phase-contrast microscopy.' *Harvard Univ. Bot. Mus. Leafl.* 21, 37–57.

JACOB, T. (1973). 'Paleoanthropological discoveries in Indonesia with special reference to the finds of the last two decades.' *J. hum. Evol.* 2, 473–485.

JACOBS, M. (1958). 'Contribution to the botany of Mount Kerintji and adjacent area in West Central Sumatra - 1.' *Ann. bogor.* 3, 45–104.

JACOBSON, G. (1970). 'Gunong Kinabalu area, Sabah, Malaysia.' *Geol. Survey Malaysia, Report,* 8, 111pp, Govt. Printer, Kuching.

JANZEN, D. H. (1970). 'Herbivores and the number of tree species in tropical forests.' *Am. Nat.* 104, 501–528.

JARDINÉ, S. (1974). 'Microflores des formations du Gabon attribuées au Karroo.' *Rev. Palaeobot. & Palynol.* 17, 75–112.

JARDINÉ, S. and MAGLOIRE, L. (1965). 'Palynologie et stratigraphie du Crétacé des Bassins du Sénégal et du Côte d'Ivoire.' *Mem. du Bureau de Recherches Geologiques et Minières* **32**, 187–245.

JONES, E. W. (1956). 'Ecological studies on the rain forest of southern Nigeria. IV. The plateau forest of the Okomu Forest Reserve (contd.).' *J. Ecol.* **44**, 83–117.

JONES, R. (1975). 'The Neolithic, Palaeolithic and the Hunting Gardeners: Man and Land in the Antipodes.' *Bull. R. Soc. N.Z.* **13**, 21–34.

KEAST, A. (1959). 'The Australian Environment.' *Biogeography and Ecology in Australia. Monographiae Biol.* **8**, (ed. A. Keast, R. L. Crocker and C. S. Christian) Chapter 2, W. Junk, The Hague.

KEAY, R. W. J. (1955). 'Montane vegetation and flora in the British Cameroons.' *Proc. Linn. Soc. Lond.* **165** (2), 140–143.

KELLMAN, M. C. (1970). *Secondary plant succession in Tropical Montane Mindanao*, 174pp, Australian National University, Research School of Pacific Studies, Department of Biogeography & Geomorphology Publication BG/2. Canberra.

KELLMAN, M. C. (1975). 'Evidence for Late Glacial Age fire in a tropical montane savanna.' *J. Biogeogr.* **2**, 57–63.

KENDALL, R. L. (1969). 'An ecological history of the Lake Victoria Basin.' *Ecol. Monogr.* **39**, 121–176.

KERSHAW, A. P. (1970a). 'A pollen diagram from Lake Euramoo, north-east Queensland.' *New Phytol.* **69**, 785–805.

KERSHAW, A. P. (1970b). 'Pollen morphological variation within the Casuarinaceae.' *Pollen Spores* **12**, 145–161.

KERSHAW, A. P. (1971). 'A pollen diagram from Quincan Crater, northeast Queensland, Australia.' *New Phytol.* **70**, 669–681.

KERSHAW, A. P. (1973a). *Late Quaternary Vegetation of the Atherton Tableland, North-East Queensland, Australia.* Ph.D thesis, Australian National University, Canberra. 439pp.

KERSHAW, A. P. (1973b). 'The numerical analysis of modern pollen spectra from northeast Queensland rainforest.' *Mesozoic & Cainozoic Palynology: Essays in Honor of Isabel Cookson*, (ed. J. E. Glover & G. Playford) 191–199, *Spec. Publ. geol. Soc. Aust.* **4**.

KERSHAW, A. P. (1975a). 'Stratigraphy and pollen analysis of Bromfield Swamp, North Eastern Queensland, Australia.' *New Phytol.* **75**, 173–191.

KERSHAW, A. P. (1975b). 'Late Quaternary Vegetation and Climate in Northeastern Australia.' *Bull. R. Soc. N.Z.* **13**, 181–187.

KERSHAW, A. P. (1976). 'A Late Pleistocene and Holocene pollen diagram from Lynch's Crater, North-Eastern Queensland, Australia.' *New Phytol.* **77**, 469–498.

KERSHAW, A. P. (in press). 'The history of vegetation and climate through at least the last 100 000 years on the Atherton Tableland, North Eastern Australia.' *Proc. IV International Palynological Conference, Lucknow, Jan. 1977.*

KERSHAW, A. P. and HYLAND, B. P. M. (1975). 'Pollen transfer and periodicity in a rain-forest situation.' *Rev. Palaeobot. & Palynol.* **19**, 129–138.

KHAN, A. M. (1974). 'Palynology of Neogene sediments from Papua (New Guinea). Stratigraphic boundaries.' *Pollen Spores* **16**, 265–284.

KINGDON, J. (1971). *East African Mammals. An atlas of evolution in Africa.* Vol. 1. Academic Press. London & New York.

KOCHUMMEN, K. M. (1966). 'Natural plant succession after farming in Sungei Kroh.' *Malay. Forester* **29**, 170–181.

KOENIGUER, J. C. (1971). 'Sur les bois fossiles du Paleocene de Sessao (Niger).' *Rev. Palaeobot. & Palynol.* **12**, 303–323.

KRAPOVICKAS, A. (1969). 'The origin, variability and spread of the groundnut (*Arachis hypogaea*).' *The domestication and exploitation of plants and animals*, (ed. P. J. Ucko and G. W. Dimbleby) 427–441, Duckworth, London.

KRÄUSEL, R. (1923). '*Nipadites borneensis* n. sp., eine fossile

Palmenfrucht aus Borneo.' *Senckenbergiana* **5** (iii–iv), 77–81.

KRÄUSEL, R. (1929). 'Fossile Pflanzen aus dem Tertiär von Süd-Sumatra.' *Verhand. Geol-Mijnb. Genootschap Nederlanden Koloniën (Geol. Ser.)* **9**, 335–378.

KRUTZSCH, W. (1967). *Atlas der mittel und jungtertiaren dispersen Sporen und Pollen-sowie der Mikroplanktonformer des nordlichen Mittel-europas.* Publ. VEB Deutscher Verlag der Wissenschaften (Vols. I-III), and Gustav Fischer Verlag (Vols. IV–VII). Berlin.

LABOURIAU, L. G. (Ed.) (1966). Segundo Simposio sobre o Cerrado. *An. Acad. brasil. Cienc.* **38** (Suppl.), 1–346.

LAKHANPAL, R. N. (1970). 'Tertiary floras of India and their bearing on the historical geology of the region.' *Taxon* **19**, 675–694.

LAM, H. J. (1945). 'Fragmenta Papuana I-VII.' Eng. Trans. in *Sargentia* **5**, 1–196.

LANGDALE BROWN, I., OSMASTON, H. A. and WILSON, J. G. (1964). *The vegetation of Uganda and its bearing on land use*, 159pp. Govt. Printer, Entebbe.

LAURENT, R. F. (1973). 'A parallel survey of Equatorial Amphibians and Reptiles in Africa and South America.' *Tropical Forest Ecosystems in Africa and South America: a comparative review*, (ed. B. J. Meggers, E. S. Ayensu and W. D. Duckworth) 259–266, Smithsonian Institution Press, Washington.

LESTREL, P. E. (1976). 'Hominid Brain Size versus Time: Revised Regression Estimates.' *J. hum. Evol.* **5**, 207–212.

LEWIS, D. (1972). *We, the navigators: the ancient art of landfinding in the Pacific*, 348pp, Australian National University Press, Canberra.

LINARES, O. F., SHEETS, D. P. and ROSENTHAL, E. J. (1975). 'Prehistoric Agriculture in Tropical Highlands.' *Science, N. Y.* **187**, 137–145.

LIND, E. M. and MORRISON, M. E. S. (1974). *East African vegetation*, 257pp. Longman, London.

LIVINGSTONE, D. A. (1962). 'Age of deglaciation in the Ruwenzori Range, Uganda.' *Nature, Lond.* **194**, 859–860.

LIVINGSTONE, D. A. (1967). 'Postglacial vegetation of the Ruwenzori Mountains in equatorial Africa.' *Ecol. Monogr.* **37**, 25–52.

LIVINGSTONE, D. A. (1971a). 'Speculations on the climatic history of mankind.' *Am. Scient.* **59**, 332–337.

LIVINGSTONE, D. A. (1971b). 'A 22 000-year pollen record from the plateau of Zambia.' *Limnol. Oceanogr.* **16**, 349–356.

LIVINGSTONE, D. A. (1975). 'Late Quaternary climatic change in Africa.' *Annu. Rev. Ecol. & Syst.* **6**, 249–280.

LONGMAN, J. A. and JENIK, J. (1974). *Tropical Forest and its Environment*, 196 pp, Longman, London.

LOUVET, P. (1973). 'Sur les affinités des flores tropicales ligneuses africaines tertiare et actuelle.' *Bull. Soc. bot. Fr.* **120**, 385–395.

LYELL, C. (1850). *Principles of Geology*, 8th Edn. 811pp, Murray, London.

LYNCH, T. F. (1974). 'The antiquity of man in South America.' *Quaternary Res.* **4**, 356–377.

MABBERLEY, D. J. (1975). 'The giant lobelias: Toxicity, inflorescence and tree-building in the Campanulaceae.' *New Phytol.* **75**, 289–295.

MacARTHUR, R. H. (1955). 'Fluctuations of animal populations and a measure of community stability.' *Ecology* **36**, 533–536.

MacARTHUR, R. H. and WILSON, E. O. (1967). *The theory of island biogeography*, 203pp. Princeton University Press, Princeton, New Jersey.

MacGINITIE, H. D. (1937). 'The flora of the Weaverville beds, of Trinity County, California.' *Publs Carnegie Instn* **465**, 83–151.

MacGINITIE, H. D. (1953). 'Fossil plants of the Florissant beds, Colorado.' *Publs Carnegie Instn* **599**, 198.

MALDONADO-KOERDELL, M. (1964). 'Geohistory and paleogeography of Middle America.' *Handbook of Middle American Indians*, (ed. R. Wauchope and R. C. West) Vol I, 3–32, University of Texas Press, Austin.

MALEY, J. (1972). 'La sédimentation pollinique actuelle dans la zone du lac Tchad (Afrique centrale).' *Pollen Spores* 14, 263–307.

MALEY, J. (1973). 'Les variations climatiques dans le Bassin du Tchad durant le dernier Millénaire: Nouvelles données palynologiques et paléoclimatiques.' *Le Quaternaire. Geodynamique, stratigraphie et environnement. Travaux Français Recents.*, 175–181. Comité National Français de l'INQUA. Paris.

MALONEY, B. K. (1975). 'Hydroseral Changes at Pea Sim-sim Swamp, North Sumatra.' *Sumatra Research Bulletin* 4, No 2, 28–37.

MALONEY, B. K. (in prep.). *Man's influence on the vegetation of North Sumatra: a palynological study*. Ph.D. thesis, University of Hull, Hull, England.

MANGELSDORF, P. C., MacNEISH, R. S. and GALINAT, W. C. (1964). 'Domestication of corn.' *Science, N. Y.* 143, 538–545.

MARSHALL, A. G. (1973). 'Conservation in West Malaysia: the Potential for International Cooperation.' *Biol. Conserv.* 5, 133–140.

MARTIN, P. S. (1964). 'Paleoclimatology and a Tropical Pollen Profile.' *Report of the VIth International Congress on Quaternary, Warsaw 1961*. Vol. ii, 319–323; Palaeoclimatological Section, Lodz. (Also Contribution 46, Program in Geochronology, University of Arizona, Tucson).

MARTIN, P. S. (1973). 'The discovery of America.' *Science, N. Y.* 179, 969–974.

MASSART, J., BOUILLENS, R., LEDOUX, P., BRIEN, P. and NAVEZ, A. (1929). *Une Mission Biologique Belge au Bresil*, vol. 2, 67pp., Imp. Medicale et Scientifique, Bruxelles, Belgique.

MAYR, E. (1944). 'Wallace's line in the light of recent zoogeographic studies.' *Q. Rev. Biol.* 19, 1–14.

McELHINNY, M. W., HAILE, N. S. and CRAWFORD, A. R. (1974). 'Palaeomagnetic evidence shows Malay Peninsula was not part of Gondwaland.' *Nature, Lond.*, 252, 641–645.

MEDWAY, Lord (1972). 'The Quaternary mammals of Malesia: a review.' *The Quaternary Era in Malesia, Transactions of the Second Aberdeen-Hull Symposium on Malesian Ecology, Aberdeen 1971*, (ed. P. and M. Ashton) 63–83, University of Hull, Dept. of Geography, Miscellaneous Series No 13, Hull, England.

MEGGERS, B. J., AYENSU, E. S. and DUCKWORTH, W. D. (Eds.) (1973). *Tropical forest ecosystems in Africa and South America: A Comparative Review*. Smithsonian Institution Press, Washington.

MOAR, N. T. (1969). 'Possible long-distance transport of pollen to New Zealand.' *N. Z. J. Bot.* 7, 424–426.

MOHR, E. C. J. and BAREN, F. A. van (1960). *Tropical Soils: a critical study of soil genesis as related to climate, rock and vegetation*, 498pp, A. Manteau, Bruxelles.

MOREAU, R. E. (1966). *The Bird Faunas of Africa and its Islands*, 424pp, Academic Press, New York.

MORLEY, R. J. (1976). *Vegetation change in West Malesia during the Late Quaternary Period*. Ph.D. thesis, University of Hull, Hull, England, 505pp. (Vol I) and Plates (Vol II).

MORLEY, R. J. (in press). 'Changes of dry-land vegetation in the Kerinci area of Central Sumatra during the Late Quaternary Period.' *Proc. IV International Palynological conference, Lucknow, Jan. 1977*.

MORLEY, R. J., FLENLEY, J. R. and KARDIN, M. K. (1973). 'Preliminary notes on the stratigraphy and vegetation of the swamps and small lakes of the Central Sumatran Highlands.' *Sumatra Research Bulletin* 2 (2), 50–60.

MORRISON, M. E. S. (1961). 'Pollen analysis in Uganda.' *Nature, Lond.* 190, 483–486.

MORRISON, M. E. S. (1966). 'Low-latitude vegetation history with special reference to Africa.' *Proceedings of the International Symposium on World Climate from 8000 to 0 B.C. London, April 1966.* 142–148, Royal Meteorological Society, London.

MORRISON, M. E. S. (1968). 'Vegetation and climate in the uplands of south-western Uganda during the later Pleistocene Period. I. Muchoya Swamp, Kigezi District.' *J. Ecol.* 56, 363–384.

MORRISON, M. E. S. and HAMILTON, A. C. (1974). 'Vegetation and climate in the uplands of south-western Uganda during the Later Pleistocene Period. II. Forest clearance and other vegetational changes in the Rukiga Highlands during the past 8000 years.' *J. Ecol.* 62, 1–31.

MORTON, J. K. (1962). 'The upland floras of West Africa—their composition, distribution and significance in relation to climatic changes.' *Compt. Rend. 4th Réunion AETFAT, Lisbon*, 391–409.

MOSELEY, M. E. (1975). *The maritime foundations of Andean civilization*, Cummings Publ. Co., Inc., Menlo Park, California.

MULLER, J. (1959). 'Palynology of Recent Orinoco delta and shelf sediments.' (Reports of the Orinoco Shelf Expedition, Vol. 5.) *Micropaleontology* 5, 1–32.

MULLER, J. (1964). 'A Palynological Contribution to the History of the Mangrove Vegetation in Borneo.' *Ancient Pacific Floras - The Pollen Story*, (ed. L. M. Cranwell) 33–42, *10th Pacific Science Congress Series*, University of Hawaii Press, Honolulu.

MULLER, J. (1965). 'Palynological study of Holocene peat in Sarawak.' *Symposium on ecological research in humid tropics vegetation, Kuching, Sarawak, July 1963*, 147–156, UNESCO.

MULLER, J. (1966). 'Montane pollen from the Tertiary of N. W. Borneo.' *Blumea* 14, 231–235.

MULLER, J. (1968). 'Palynology of the Pedawan and Plateau Sandstone Formations (Cretaceous-Eocene) in Sarawak, Malaysia.' *Micropaleontology* 14, 1–37.

MULLER, J. (1970). 'Palynological evidence on early differentiation of angiosperms.' *Biol. Rev.* 45, 417–450.

MULLER, J. (1972). 'Palynological evidence for change in geomorphology, climate and vegetation in the Mio-Pliocene of Malesia.' *The Quaternary Era in Malesia, Transactions of the Second Aberdeen-Hull Symposium on Malesian Ecology, Aberdeen 1971*, (ed. P. and M. Ashton) 6–16, University of Hull, Dept. of Geography, Miscellaneous Series No 13, Hull, England.

MULVANEY, D. J. and SOEJONO, R. P. (1970). 'Archaeology in Sulawesi, Indonesia.' *Antiquity* 45, 26–33.

NELSON, H. E. (1971). 'Disease, demography and the evolution of social structure in the New Guinea Highlands.' *J. Polynes. Soc.* 80, 204–216.

NEWELL, R. E. (1973). 'Climate and the Galapagos Islands.' *Nature, Lond.* 245, 91–92.

NIX, H. A. and KALMA, J. D. (1972). 'Climate as a dominant control in the biogeography of northern Australia and New Guinea.' *Bridge and Barrier: The Natural and Cultural History of Torres Strait*, (ed. D. Walker), 61–92, Australian National University, Research School of Pacific Studies, Dept. of Biogeography and Geomorphology, Publ. BG/3. Canberra.

NOSSIN, J. J. (1962). 'Coastal sedimentation in north eastern Johore (Malaya).' *Z. Geomorph.* 6, 296–316.

OLLIER, C. D. (1974). 'Jungles and Rain Forests (part).' *Encyclopaedia Britannica* 15th Edition. 10, 336–342.

OSMASTON, H. A. (1958). *Pollen analysis in the study of the past vegetation and climate of Ruwenzori and its neighbourhood*, B. Sc. thesis, 44pp, Oxford University, Oxford, England.

OSMASTON, H. A. (1965). *The past and present climate and vegetation of Ruwenzori and its neighbourhood*, Ph.D. thesis, Oxford University, Oxford, England.

PALMER, P., LIVINGSTONE, D. and KINGSOLVER, J. (in press). 'New approaches to microfossil identification in the highlands of equatorial Africa.' *Proc. IV International Palynological Conference, Lucknow, Jan. 1977.*

PATTERSON, T. C. (1971). 'The emergence of food production in Central Peru.' *Prehistoric Agriculture,* (ed. S. Struever) 181–207, Natural History Press, New York.

PEARSON, R. (1964). *Animals and plants of the Cenozoic Era,* 236pp, Butterworth, London.

PERCIVAL, M. (1950). 'Pollen presentation and pollen collection.' *New Phytol.* **49,** 40–63.

PERERA, N. P. (1968). 'Some problems of Climate-Vegetation Correlations with special reference to Ceylon.' *Vidyodaya J. Arts, Sci., Lett.* **1**(2) 173–184.

PETROSYANTS, M. A. and TROFIMOV, D. M. (1971). 'Sporova - Piltsevaya Characteristica Verugnemelorik Otlovenii Mali-Nigerskoi.' *Bulletin M. 0-VA Prirody Otd. Geologii,* **46** (6), 75– (in Russian).

PETROV, SI. and DRAZHEVA-STAMATOVA, TS. (1972). '*Reevesia* Lindl. fossil pollen in the Tertiary sediments of Europe and Asia.' *Pollen Spores* **14,** 79–95.

PICKERSGILL, B. (1969). 'The domestication of chili peppers.' *The domestication and exploitation of plants and animals,* (ed. P. J. Ucko and G. W. Dimbleby) 443–450, Duckworth, London.

POLAK, E. (1933). 'Ueber Torf und Moor in Niederlandisch Indien.' *Proc. K. ned. Akad. Wet.* **30,** 1–84.

POLAK, E. (1949). 'De Rawa Lakbok, een eutroof laagveen op Java.' *Meded. Alg. Proefstn Landb., Buitenz.* **85,** 1–60.

POLAK, E. (1951). 'Construction and origin of floating islands in the Rawa Pening (Central Java).' *Contr. gen. agric. Res. Stn Bogor* **121,** 1–13.

POORE, M. E. D. (1964). 'Integration in the plant community.' *J. Ecol.* **52** (suppl), 213–226.

POSNANSKY, M. (1967). 'The Iron Age in East Africa.' *Background to Evolution in Africa,* (ed. W. W. Bishop and J. D. Clark) 629–649, University of Chicago Press, Chicago.

POST, L. von (1916). 'Om skogsträdpollen i sydsvenska torfmosselagerföljder.' *Geol. För. Stockh. Förh.* **38,** 384–394.

POST, L. von (1946). 'The prospect for pollen analysis in the study of the earth's climatic history.' *New Phytol.* **45,** 193–217.

POSTHUMUS, O. (1931). 'Plantae.' *Festbundel K. Martin, Leid. geol. Meded.* **5,** 485–508.

POTBURY, S. S. (1935). 'The La Porte flora of Plumas County, California.' *Publs Carnegie Instn* **465,** 29–81.

POTTER, G. L., ELLSAESSER, H. W., MacCRACKEN, M. C. and LUTHER, F. N. (1975). 'Possible climatic impact of tropical deforestation.' *Nature, Lond.* **258,** 697–698.

POWELL, C. McA. and CONAGHAN, P. J. (1973). 'Plate tectonics and the Himalayas.' *Earth Planet. Sci. Lett.* **20,** 1–12.

POWELL, J. M. (1970). *The impact of man on the vegetation of the Mt. Hagen Region, New Guinea.* Unpub. Ph.D. thesis, Australian National University, Canberra, 218pp and appendices.

POWELL, J. M. (1976). 'Part III Ethnobotany,' *New Guinean Vegetation,* (ed. K. Paijmans) 106–183 Australian National University Press, Canberra.

POWELL, J. M. (in press a). 'The origins of agriculture in New Guinea.' *Proc. IV International Palynological Congress, Lucknow, Jan. 1977.*

POWELL, J. M. (in press b). 'Studies in New Guinea Vegetation History.' *Proc. IV International Palynological Conference, Lucknow, Jan. 1977.*

POWELL, J. M., KULUNGA, A., MOGE, R., PONO, C., ZIMIKE, F. and GOLSON, J. (1975). *Agricultural traditions of the Mount Hagen Area.* University of Papua New Guinea, Department of Geography, Occasional Paper No 12, 67pp and 5 figs.

PRAKASH, U. (1965). 'A survey of the fossil dicotyledonous woods from India and the Far East.' *J. Paleont.* **39,** 815–827.

PRAKASH, U. (1973). 'Palaeoenvironmental analysis of Indian Tertiary floras.' *Geophytology* **2,** 178–205.

PRANCE, G. T. (1973). 'Phytogeographic support for the theory of Pleistocene forest refuges in the Amazon Basin, based on evidence from distribution patterns in Caryocaraceae, Chrysobalanaceae, Dichapetalaceae and Lecythidaceae.' *Acta Amazonica* **3** (3), 5–28.

RAUP, H. M. (1964). 'Some problems in ecological theory and their relation to conservation.' *J. Ecol.* **52** (suppl), 19–28.

RAVEN, P. H. and AXELROD, D. I. (1974). 'Angiosperm biogeography and past continental movements.' *Ann. Mo. bot. Gdn.* **61** (3), 539–673.

REGALI, M. da S. P., UESUGUI, N. and SANTOS, A. da S. (1974). 'Palinologia dos sedimentos meso-cenozoicos do Brasil (I).' *B. tec. Petrobras, Rio de Janeiro,* **17** (3), 177–191.

REID, E. M. and CHANDLER, M. E. J. (1933). *The Flora of the London Clay,* 561pp, 33 pls, Brit. Mus. (Nat. Hist.), London.

RICHARDS, P. W. (1939). 'Ecological studies on the rain forest of Southern Nigeria. I. The structure and floristic composition of the primary forest.' *J. Ecol.* **27,** 1–61.

RICHARDS, P. W. (1955). 'The secondary succession in the tropical rain forest.' *Sci. Prog., Lond.* **43,** 45–57.

RICHARDS, P. W. (1964). *The Tropical Rain Forest. An Ecological Study,* 2nd Reprint, 450pp, Cambridge University Press, Cambridge.

RICHARDS, P. W. (1973). 'Africa, the "Odd Man Out".' *Tropical Forest Ecosystems in Africa and South America: a comparative review.* (Ed. B. J. Meggers, E. S. Ayensu and W. D. Duckworth) 21–26, Smithsonian Institution Press, Washington.

RICHARDSON, J. A. (1947). 'An outline of the geomorphological evolution of British Malaya.' *Geol. Mag.* **84,** 129–144.

RICHARDSON, J. L. and RICHARDSON, A. E. (1972). 'History of an African Rift Lake and its Climatic Implications.' *Ecol. Monogr.* **42,** 499–534.

RICKLEFS, R. E. (1973). *Ecology,* 861pp, Nelson, London.

RILEY, C. L., KELLEY, J. C., PENNINGTON, C. W. and RANDS, R. L. (Eds.) (1971). *Man across the sea: Problems of Pre-Columbian contacts,* 552pp, University of Texas Press, Austin, Texas, U. S. A.

ROBBINS, R. G. (1958). 'Montane Formations in the Central Highlands of New Guinea.' *Proceedings of Symposium on Humid Tropics Vegetation, Tjiawi (Indonesia), December 1958,* 176–195 UNESCO.

ROSS, R. (1954). 'Ecological Studies on the Rainforest of Southern Nigeria. III Secondary Succession in the Shasha Forest Reserve.' *J. Ecol.* **42,** 259–282.

ROYEN, P. van (1963). 'Sertulum papuanum 11. Plantaginaceae.' *Nova Guinea, Botany* **18,** 418–426.

RUDOLPH, K. and FIRBAS, F. (1926). 'Pollen analytische Untersuchung subalpiner moores des Reisengebirges.' *Ber. dt. bot. Ges.* **44,** 227–238.

SALGADO-LABOURIAU, M. L. (1973). *Contribuição á Palinologia dos Cerrados.* 291 pp, Academia Brasileira de Ciencias, Rio de Janeiro.

SALGADO-LABOURIAU, M. L. (in press). 'Holocene pollen analyses from the Venezuelan Andes.' *Proc. IV International Palynological Conference, Lucknow, Jan. 1977.*

SALGADO-LABOURIAU, M. L. and SCHUBERT, C. (1976). 'Palynology of Holocene Peat Bogs from the central Venezuelan Andes.' *Palaeogeogr., Palaeoclimatol., Palaeoecol.* **19,** 147–156.

SANBORN, E. I. (1935). 'The Comstock flora of west central Oregon,' *Publs Carnegie Instn* **465,** 1–28.

SARTONO, S. (1970). 'The discovery of a hominid skull at Sangiran, Central Java.' *Directorat Geologi special publication* No 3, 11pp. Bandung.

SAUER, C. O. (1952). *Agricultural origins and dispersals.* The American Geographical Society, New York.

SCHIMPER, A. F. W. (1903). *Plant—geography upon a physiological basis.* (Transl. by W. R. Fisher, ed. P. Groom & I. B. Balfour) 839pp, Oxford.

SCHNELL, R. (1950). *La forêt dense. Introduction à l'étude botanique de la région forestière d'Afrique occidentale.* Manuels ouest-africains, Vol. 1, (ed. P. Lechevalier) 330pp, Paris.

SCHOLL, D. W. (1968). 'Mangrove Swamps: geology and sedimentology.' *The Encyclopedia of Geomorphology* (ed. P. W. Fairbridge) 683—696, Reinhold, New York.

SCHULZ, J. P. (1960). 'Ecological Studies on Rain Forest in Northern Suriname.' *Proc. K. ned. Akad. Wet., Sect. C* 53, 1—267.

SCHUSTER, J. (1911a). 'Die Flora der Trinil-Schichten.' *Die Pithecanthropus Schichten auf Java* (by Selenka and Blanckenhorn) 235—257. Leipzig.

SCHUSTER, J. (1911b). 'Monographie der fossilen Flora der *Pithecanthropus* Schichten.' *Abh. bayer. Akad. Wiss.* 25, vi.

SEWARD, A. C. (1924). 'A collection of fossil plants from south-eastern Nigeria.' *Bull. geol. Surv. Nigeria* 6, Appendix II, 66—81.

SHACKLETON, N. (1967). 'Oxygen isotope analyses and Pleistocene temperatures reassessed.' *Nature, Lond.* 215, 15—17.

SHARMA, C. and SINGH, G. (1972a). 'Studies in the Late-Quaternary Vegetational History in Himashal Pradesh - 1. Khajiar Lake.' *Palaeobotanist* 21, 144—162.

SHARMA, C. and SINGH, G. (1972b). 'Studies in the Late-Quaternary Vegetational History in Himashal Pradesh - 2. Rewalsar Lake.' *Palaeobotanist* 21, 321—338.

SIMMONS, I. G., ATHERDEN, M. A., CUNDILL, P. R. and JONES, R. L. (1975). 'Inorganic layers in soligenous mires of the North Yorkshire Moors.' *J. Biogeogr.* 2, 49—56.

SIMPSON, B. (1971). 'Pleistocene changes in the fauna and flora of South America.' *Science, N. Y.* 173, 771—780.

SIMPSON, B. (1973). 'Pleistocene speciation in the mountains of tropical South America.' *First Int. Congr. Syst. Ecol. Biol. Boulder.*

SIMPSON, B. B. (1975). 'Glacial climates in the eastern tropical South Pacific.' *Nature, Lond.* 253, 34—36.

SINGH, G. (1971). 'The Indus Valley Culture seen in the context of post-glacial climatic and ecological studies in North-West India.' *Archaeol. & Phys. Anthropol. Oceania* 6, 177—189.

SINGH, G., JOSHI, R. D., CHOPRA, S. K. and SINGH, A. B. (1974). 'Late Quaternary history of vegetation and climate of the Rajasthan Desert, India.' *Phil. Trans. R. Soc. B.* 267, 467—501.

SMARTT, J. (1969). 'Evolution of American *Phaseolus* beans under domestication.' *The domestication and exploitation of plants and animals,* (ed. P. J. Ucko and G. W. Dimbleby) 451—462, Duckworth, London.

SMILEY, C. J. (1967). 'Palaeoclimatic interpretations of some Mesozoic floral sequences.' *Bull. Am. Ass. Petrol. Geol.* 51 (6), 849—863.

SMITH, A. G., BRIDEN, J. C. and DREWRY, G. E. (1973). 'Phanerozoic World Maps' *Organisms and Continents through time,* (ed. N. F. Hughes) 1—42, Special papers in Palaeontology, No 12.

SMITH, D. M., GRIFFIN, J. J. and GOLDBERG, E. D. (1973). 'Elemental Carbon in marine sediments: a baseline for burning.' *Nature, Lond.* 241, 268—270.

SMITH, J. M. B. (1974). *Origins and ecology of the non-forest flora of Mt Wilhelm, New Guinea.* Ph.D. thesis, Australian National University, Canberra, 270pp.

SNEATH, P. H. A. (1967). 'Conifer distributions and continental drift.' *Nature, Lond.* 215, 467—470.

SOBUR, A. S., CHAMBERS, M. J., CHAMBERS, R., DAMOPOLII, J., HADI, S. and HANSON, A. J. (1975). 'Remote sensing applications for environmental and resource management studies in the Musi Banyuasin coastal zone, South Sumatra.' *UN/FAO Regional Seminar on Remote Sensing Applications, November 1975, Jakarta, Indonesia.*

SOEPADMO, E. (1972). 'Fagaceae.' *Flora Malesiana* 7, 265—403.

SOLHEIM, W. G. (1970). 'Northern Thailand, Southeast Asia, and World Prehistory.' *Asian Perspectives,* 13, 145—162.

SPORNE, K. R. (1969). 'The ovule as an indicator of evolutionary status in angiosperms.' *New Phytol.* 68, 555—566.

SPORNE, K. R. (1971). *The mysterious origin of flowering plants.* Oxford Biology Readers, No 3. 16pp. Oxford University Press, London.

SPORNE, K. R. (1973). 'The survival of archaic dicotyledons in tropical rain-forests.' *New Phytol.* 72, 1175—1184.

STAUFFER, P. H. (1973). 'Cenozoic.' *Geology of the Malay Peninsula (West Malaysia and Singapore),* (ed. D. J. Gobbett and C. S. Hutchison) Chapter 6, Wiley-Interscience, New York.

STEENIS, C. G. G. J. van (1934—36). 'On the Origin of the Malaysian Mountain Flora.' *Bull. Jard. bot. Buitenz. Series III;* Part 1, 13, 135—262; Part II, 13, 289—417; Part III, 14, 56—72.

STEENIS, C. G. G. J. van (1935). 'Maleische Vegetatieschetsen.' *Tijdschr. K. ned. aardrijksk. Genoot.* 52, 25—67, 171—203, 363—398.

STEENIS, C. G. G. J. van (1938). 'Exploraties in de Gajo-Landen. Algemeene Resultaten der Losir-Expeditie, 1937.' *Tijdschr. K. ned. aardrijksk. Genoot.* 55, 727—801.

STEENIS, C. G. G. J. van (1958a). 'Rejuvenation as a factor for judging the status of vegetation types: the biological nomad theory.' *Proceedings of the Kandy Symposium on Humid Tropics Vegetation 1956,* 212—215, UNESCO.

STEENIS, C. G. G. J. van (1958b). *Vegetation map of Malaysia,* 1:5,000,000. Published in collaboration with UNESCO for the UNESCO humid tropics research project.

STEENIS, C. G. G. J. van (1962a). 'The Mountain Flora of the Malaysian Tropics.' *Endeavour* 21, 183—193.

STEENIS, C. G. G. J. van (1962b). 'The distribution of mangrove plant genera and its significance for palaeogeography.' *Proc. K. ned. Akad. Wet. Ser. C.* 65, 164—169.

STEENIS, C. G. G. J. van (1964). 'Plant Geography of the Mountain Flora of Mt Kinabalu.' *Proc. R. Soc. B* 161, 7—38.

STEENIS, C. G. G. J. van (1968). 'Frost in the tropics.' *Proceedings of the Symposium on Recent Advances in Tropical Ecology,* Part I, (ed. R. Misra and B. Gopal) 154—167, Shri R. K. Jain, Faridabad, India.

STEENIS, C. G. G. J. van (1971). '*Nothofagus,* key genus of plant geography, in time and space, living and fossil, ecology and phylogeny.' *Blumea* 19, 65—98.

STEENIS, C. G. G. J. van (1972). *The Mountain flora of Java.* (Illustrated by Amir Hamzah and Moehamad Toha) 90pp, E. J. Brill, Leiden.

STEENIS, C. G. G. J. van and SCHIPPERS-LAMMERTSE, A. F. (1965). 'Concise Plant-Geography of Java.' Pp 1—72 in Vol. 2 of *Flora of Java* (C. A. Backer and R. C. Bakhuizen van den Brink).

STOVER, L. E. (1964). 'Cretaceous ephedroid pollen from West Africa.' *Micropaleontology* 10, 145—156.

TANSLEY, A. G. (1920). 'The classification of vegetation and the concept of development.' *J. Ecol.* 8, 118—149.

TAUBER, H. (1967). 'Investigations of the mode of pollen transfer in forested areas.' *Rev. Palaeobot. & Palynol.* 3, 277—286.

TAYLOR, B. W. (1957). 'Plant succession on recent volcanoes in Papua.' *J. Ecol.* 45, 233—243.

THANIKAIMONI, G. (1968). 'Palynology in Vegetation Studies.' *Symposium on phytogeography and vegetation mapping. 21st Int. Geographical Congress 1968.* Calcutta, 2—4.

THOM, B. G. (1967). 'Mangrove ecology and deltaic geomorphology: Tabasco, Mexico.' *J. Ecol.* 55, 301—343.

THOMAS, W. L., SAUER, C. O., BATES, M. and MUMFORD, L., (Eds.) (1956). *Man's Role in Changing the Face of the Earth,* 1193pp. University of Chicago Press, Chicago.

TOMLINSON, R. W. (1974). 'Preliminary Biogeographical Studies on the Inyanga Mountains, Rhodesia.' *S. Afr. geogr. J.* 56, 15—26.

TRALAU, H. (1964). 'The genus *Nypa* van Wurmb.' *K. svenska Vetensk-Akad. Handl.* **10**(1), 1–29.

TRICART, J. (1974). 'Existence de Périodes Sèches au Quaternaire en Amazonie et dans le régions voisines.' *Revue Géomorph. dyn.* **23**, 145–158.

TROLL, C. (1959). *Die tropischen Gebirge. Ihre dreidimensionale klimatische und pflanzengeographische Zonierung,* 93pp, Dummlers, Bonn.

TSCHUDY, R. H. and SCOTT, R. A. (eds) (1969). *Aspects of Palynology. An introduction to plant microfossils in time*, 510pp, Wiley-Interscience, New York.

TSUKADA, M. (1966). 'Late Pleistocene vegetation and climate in Taiwan (Formosa)'. *Proc. natn. Acad. Sci. U.S.A.* **55**, 543–548.

TSUKADA, M. (1967). 'Vegetation in subtropical Formosa during the Pleistocene glaciations and the Holocene.' *Palaeogeogr. Palaeoclimatol. Palaeoecol.* **3**, 49–64.

TSUKADA, M. and DEEVEY, E. S. (1967). 'Pollen analyses from four lakes in the Southern Maya Area of Guatemala and El Salvador.' *Quaternary Paleoecology,* (ed. E. J. Cushing and H. E. Wright, Jr) 303–331, Yale Univeristy Press, New Haven and London.

TSUKADA, M. and ROWLEY, J. R. (1964). 'Identification of modern and fossil maize pollen.' *Grana palynol.* **5**, 406–412.

UNITED NATIONS (1971). *United Nations list of national parks and equivalent reserves.* 2nd Edn. 601pp, Heyez, Brussels.

VANN, J. H. (1969). *The physical geography of the lower coastal plain of the Guiana coast.* Geogr. Branch, ONR, Washington, D. C. Project NR 388–028, Tech. Rept. 1, 91pp.

VANZOLINI, P. E. and WILLIAMS, E. E. (1970). 'Southamerican anoles: the geographic differentiation and evolution of the *Anolis chrysolepis* species group (Sauria, Iguanidae).' *Archos Zool. Est. S Paulo.* **19**, 1–124.

VAVILOV, N. I. (trans. K. Starr Chester) (1951). *The origin, variation, immunity and breeding of cultivated plants.* *Chronica bot.* **13**, 1–364.

VERSTAPPEN, H. Th. (1973). *A geomorphological reconnaissance of Sumatra and adjacent islands (Indonesia),* 194pp, Wolters–Noordhoff Groningen, Netherlands.

VERSTAPPEN, H. Th. (1975). 'On palaeo-climates and landform development in Malesia.' *Modern Quaternary Research in Southeast Asia,* (ed. G. -J. Bartstra and W. A. Casparie), 3–35, Balkema, Rotterdam.

VISHNU-MITTRE (1966). 'Kaundinyapur plant economy in protohistoric and historic times.' *Palaeobotanist* **15**, 152–156.

VISHNU-MITTRE (1972). 'Neolithic plant economy at Chirand, Bihar.' *Palaeobotanist* **21**, 18–22.

VISHNU-MITTRE and GUPTA, H. P. (1971). 'The origin of Shola Forest in the Nilgiris, South India.' *Palaeobotanist* **19**, 110–114.

VOOGD, C. N. A. de (1940). 'De Batoer op Bali.' *Trop. Natuur* **29**, 37–53.

WADDELL, E. W. (1972). *The mound builders: agricultural practices, environment and society in the Central Highlands of New Guinea. Monographs of the American Ethnological Society,* **53**, University of Washington Press, Seattle.

WADE, L. K. and McVEAN, D. N. (1969). *Mt Wilhelm Studies I. The alpine and subalpine vegetation,* Australian National University, Research School of Pacific Studies, Dept. of Biogeography and Geomorphology, Publication BG/1, 225pp, Canberra.

WALKER, D. (1966). 'Vegetation of the Lake Ipea region, New Guinea Highlands, 1, Forest, grassland and "garden".' *J. Ecol.* **54**, 503–533.

WALKER, D. (1970a). 'The changing vegetation of the montane tropics.' *Search* **1**, 217–221.

WALKER, D. (1970b). 'Direction and rate in some British post-glacial hydroseres.' *Studies in the vegetational history of the British Isles. Essays in honour of Harry Godwin,* (ed. D. Walker and R. G. West) 117–139, Cambridge University Press, Cambridge.

WALKER, D. (1972a). 'Vegetation of the Lake Ipea Region, New Guinea Highlands. II Kayamanda Swamp.' *J. Ecol.* **60**, 479–504.

WALKER, D. (ed.) (1972b). *Bridge and Barrier: the Natural and Cultural History of Torres Strait.* Australian National University, Research School of Pacific Studies, Dept. of Biogeography and Geomorphology, Publication BG/3, 437pp, Canberra.

WALKER, D. and FLENLEY, J. R. (in press). 'Late Quaternary Vegetational History of the Enga District of Upland Papua New Guinea.' *Phil. Trans. R. Soc. B.*

WALKER, D. and GUPPY, J. C. (1976). 'Generic plant assemblages in the highland forests of Papua New Guinea.' *Australian Journal of Ecology* **1**, 203–212.

WALLACE, A. R. (1869). *The Malay Archipelago.* Vol. 1 pp 478, Vol. II pp 524, Macmillan, London.

WALTER, H. (1973). *Vegetation of the Earth in relation to climate and the eco-physiological conditions.* (Translated from the 2nd German edition by Joy Wieser) 237pp. English Universities Press, London.

WASHBOURN, C. K. (1967). 'Lake levels and Quaternary Climates in the Eastern Rift Valley of Kenya.' *Nature, Lond.* **216**, 672–673.

WATSON, J. B. (1965a). 'From hunting to horticulture in the New Guinea Highlands.' *Ethnology* **4**, 295–309.

WATSON, J. B. (1965b). 'The significance of a recent ecological change in the Central Highlands of New Guinea.' *J. Polynes. Soc.* **74**, 438–450.

WATSON, J. G. (1928). 'Mangrove Forests of the Malay Peninsula.' *Malay Forest Rec.* **6**, 275pp.

WATTS, D. (1970). 'Persistence and change in the vegetation of oceanic islands: an example from Barbados, West Indies.' *Can. Geogr.* **14**, 91–109.

WATTS, D. (1978). 'The new biogeography and its niche in physical geography.' *Geography,* **63**, 324–337.

WEBB, L. J. (1959). 'A physiognomic classification of Australian rain forests.' *J. Ecol.* **47**, 551–570.

WEBB, L. J. (1968). 'Environmental relationships of the structural types of Australian rain forest vegetation.' *Ecology* **49**, 296–311.

WEBB, L. J. and TRACEY, J. G. (1972). 'An ecological comparison of vegetation communities on each side of Torres Strait.' *Bridge and Barrier: the Natural and Cultural History of Torres Strait,* (ed. D. Walker) 109–129, Australian National University, Research School of Pacific Studies, Department of Biogeography and Geomorphology, Publication BG/3, Canberra.

WEBB, L. J., TRACEY, J. G. and HAYDOCK, K. P. (1967a). 'A factor toxic to seedlings of the same species associated with living roots of the non-gregarious subtropical rain forest tree *Grevillea robusta.' J. appl. Ecol.* **4**, 13–25.

WEBB, L. J., TRACEY, J. G. and WILLIAMS, W. T. (1972). 'Regeneration and pattern in the subtropical rain forest.' *J. Ecol.* **60**, 675–695.

WEBB, L. J., TRACEY, J. G., WILLIAMS, W. T. and LANCE, G. N. (1967b). 'Studies in the numerical analysis of complex rain-forest communities. 1. Comparison of methods applicable to site/species data.' *J. Ecol.* **55**, 171–191.

WEBB, L. J., TRACEY, J. G., WILLIAMS, W. T. and LANCE, G. N. (1967c). 'Studies in the numerical analysis of complex rain-forest communities. II. The problem of species sampling.' *J. Ecol.* **55**, 525–538.

WEBERBAUER, A. (1914). 'Die Vegetationsgliederung des nordlichen Peru um 5° sudl.' *Br. Bot. Jb.* **50**, 72–94.

WEBERBAUER, A. (1922). 'Die Vegetationskarte der peruanischen Anden zwischen 5° und 17°S.' *Petermanns Mitt.* **68**, 89–91 and 120–122.

WEBSTER, P. J. and STRETEN, N. A. (1972). 'Aspects of late Quaternary climate in tropical Australasia.' *Bridge and Barrier: The Natural and Cultural History of Torres*

Strait, (ed. D. Walker) 39–60, Australian National University, Research School of Pacific Studies, Dept of Biogeography and Geomorphology, Publ. BG/3, Canberra.

WEST, R. G. (1977). *Pleistocene Geology and Biology,* 2nd Edition, Longman, London.

WESTING, A. H. (1972). 'Herbicides in War: Current Status and Future Doubt.' *Biol. Conserv.* 4, 322–327.

WHITE, K. J. (1975). *The Effect of Natural Phenomena on the Forest Environment.* Presidential address to Papua New Guinea Scientific Society, 26th March 1975, Department of Forests, Port Moresby.

WHITMORE, T. C. (1974). 'Change with Time and the Role of Cyclones in Tropical Rain Forest on Kolombangara, Solomon Islands,' 92pp, *Comm. For. Inst. Paper* No 46, Oxford University Press, Oxford.

WHITMORE, T. C. (1975). *Tropical Rain Forests of the Far East,* 282pp, Oxford University Press, Oxford.

WHITTAKER, R. H. (1953). 'A consideration of climax theory: the climax as a population and pattern.' *Ecol. Monogr.* 23, 41–78.

WHYTE, R. O. (1972). 'The Gramineae, Wild and Cultivated, of Monsoonal and Equatorial Asia. I. Southeast Asia.' *Asian Perspectives* 15, 127–151.

WIJMSTRA, T. A. (1969). 'Palynology of the Alliance Well.' *Geologie Mijnb.* 48, 125–133.

WIJMSTRA, T. A. (1971). *The palynology of the Guyana coastal basin.* Dissertation, University of Amsterdam.

WIJMSTRA, T. A. and HAMMEN, T. van der (1966). 'Palynological data on the history of tropical savannas in Northern South America.' *Leid. geol. Meded.* 38, 71–90.

WILLIAMS, P. W., McDOUGALL, I. and POWELL, J. M. (1972). 'Aspects of the Quaternary Geology of the Tari-Koroba area, Papua.' *J. geol. Soc. Aust.* 18, 333–347.

WILLIAMS, W. T., LANCE, G. N., WEBB, L. J., TRACEY, J. G. and DALE, M. B. (1969). 'Studies in the numerical analysis of complex rain-forest communities. III The analysis of successional data.' *J. Ecol.* 57, 515–535.

WILLIS, J. C. (1922). *Age and Area: a study in geographical distribution and origin of species* (with chapters by H. de Vries and others) Cambridge University Press, Cambridge.

WINTERBOTTOM, J. M. (1967). 'Climatological implications of avifaunal resemblances between S. W. Africa and Somaliland.' *Palaeoecol. Af. & Surround. Isles Antarct.* 2, 77–79.

WOLFE, J. A. (1971). 'Tertiary climatic fluctuations and methods of analysis of Tertiary floras.' *Palaeogeogr. Palaeoclimatol. Palaeoecol.* 9, 27–57.

WOLFE, J. A. (1975). 'Some Aspects of Plant Geography of the Northern Hemisphere during the Late Cretaceous and Tertiary.' *Ann. Mo. bot. Gdn* 62, 264–279.

WOLFE, J. A., DOYLE, J. A. and PAGE, V. M. (1975). 'The Bases of Angiosperm Phylogeny: Palaeobotany.' *Ann. Mo. bot. Gdn* 62, 801–824.

WOLFE, J. A. and HOPKINS, D. M. (1967). 'Climatic changes recorded by Tertiary land floras in northwestern North America.' *Tertiary Correlations and Climatic Changes in the Pacific,* 67–76, Symposium 25 of the 11th Pacific Science Congress, Tokyo 1966.

WOOD, D. (1970). 'The tropical forest and Sporne's advancement index.' *New Phytol.* 69, 113–115.

WOODRING, W. P. (1966). 'The Panama land bridge as a sea barrier.' *Proc. Am. phil. Soc.* 110, 425–433.

WYATT-SMITH, J. (1966). 'Ecological studies on Malayan forests. I.' *Malayan Forestry Department Research Pamphlet* 52.

YEN, D. E. (1963). 'Sweet potato variation and its relation to human migration in the Pacific. *Plants and the migrations of Pacific peoples. Symposium of the 10th Pacific Science Congress, Hawaii, 1961,* (ed. J. Barrau) 93–117, Bishop Museum Press, Honolulu.

YEN, D. E. (1973). 'The origins of Oceanic agriculture.' *Archaeol. & Phys. Anthropol. Oceania* 8, 68–85.

YEN, D. E. (1977). 'Hoabinhian horticulture? The evidence and the question from Northwest Thailand.' *Sunda and Sahul: Prehistoric study in southeast Asia, Melanesia and Australia* (ed. J. Allen, J. Golson and R. Jones) Academic Press, London.

ZAKLINSKAYA, E. D. (1964). 'On the relationships between Upper Cretaceous and Palaeogene floras of Australia, New Zealand and Eurasia, according to data from spore and pollen analyses.' *Ancient Pacific Floras — The Pollen Story,* (ed. L. M. Cranwell) 85–86, 10th Pacific Science Congress Series, University of Hawaii Press, Honolulu.

ZUCCHI, A. (1973). 'Prehistoric human occupations of the western Venezuelan Llanos.' *Am. Antiq.* 38, 182–190.

Index

Italicized page numbers denote major sections on the topic.